Jason Hiebert
Box 152
Waldheim, Sask
S0K-4R0

MW01716201

Power Trains

JOHN DEERE

FOS
FUNDAMENTALS OF SERVICE

PUBLISHER

Fundamentals of Operation (FMO) is a series of manuals created by Deere & Company. Each book in the series is conceived, researched, outlined, edited, and published by Deere & Company. Authors are selected to provide a basic technical manuscript which is edited and rewritten by staff editors.

PUBLISHER: DEERE & COMPANY SERVICE TRAINING, Dept. F, John Deere Road, Moline, Illinois 61265; DIRECTOR OF SERVICE: Robert A. Sohl.

FUNDAMENTAL WRITING SERVICES EDITORIAL STAFF

Managing Editor: Louis R. Hathaway
Editor: Laurence T. Hammond
Editor: Alton E. O'Banion
Promotions: Annette M. LaCour

TO THE READER

PURPOSE OF THIS MANUAL

The main purpose of this manual is to help the reader understand and service power trains. Starting with "how power trains work," we build up to "why they may fail" and "what to do if they do." This manual is also an excellent reference for the trained mechanic who wants to refresh his memory on power trains. It is written in a simple form using many illustrations so that it can be understood easily.

POWER TRAINS IN THIS BOOK

"Power Trains" is a broad term. In this book, the prime interest is in transmissions used to transmit power on farm and industrial machines.

TO USE THIS BOOK

This book can be used by experienced mechanics, shop trainees, and vocational students.

By starting with the basics, we build your knowledge step by step. Chapter 1 covers the basics—"how it works;" chapter 2 covers clutches, and chapters 3 through 6 explain several kinds of transmissions. The remaining chapters cover differentials, final drives, PTO's, and special drives.

If you aren't familiar with basic power trains start at Chapter 1 and study the chapters in sequence. The experienced man can find what he needs in the table of contents.

Answers to the test yourself questions at the end of each chapter are printed at the end of the book on pages 9 and 10 following the index.

OTHER MANUALS IN THIS SERIES

- **Hydraulics**
- **Engines**
- **Electrical Systems**
- **Shop Tools**
- **Welding**
- **Air Conditioning**
- **Bearings and Seals**
- **Belts and Chains**
- **Fuels, Lubricants, and Coolants**
- **Mowing and Spraying Equipment**
- **Tires and Tracks**
- **Fiber Glass/Plastics**
- **Fasteners**
- **Identification of Parts Failures**

Each manual is backed up by a set of 35 mm color slides for classroom use.

FOR MORE INFORMATION

Write for a free *FOS Catalog of Manuals and Slides.* Send your request to:

John Deere Technical Services, Dept. F
John Deere Road
Moline, Illinois 61265

ACKNOWLEDGMENTS

John Deere gratefully acknowledges help from the following: Allision Division, General Motors Corp.; American Gear Manufacturers Assn.; American Oil Co.; American Society of Agricultural Engineers; American Society of Mechanical Engineers; Bendix Corp.; Borg-Warner Corp., Rockford Clutch Division; Cessna Aircraft Co.; Clark Equipment Co.; Dana Corporation, Spicer Axle Division; Dayco Corp.; Detroit Automotive Products Corp.; Eaton, Yale and Towne Inc., Eaton Axle Div.; Gates Rubber Co.; General Motors Corp.; Goodyear Tire and Rubber Co.; Link-Belt/FMC; Lipe-Rollway Corp.; National Fluid Power Assn.; Nuday Co.; Owatonna Tool Co.; Purolator Products Inc.; Rex Chainbelt Inc.; Rockwell-Standard Co.; Rubber Manufacturers Assn.; Society of Automotive Engineers; Sundstrand Corp.; Twin Disc Inc.; Vickers Inc., Division of Sperry Rand Corp.; Warner Electric Brake and Clutch Co.

FOS—40 Litho in U.S.A. ISBN 0-86691-084-0

CONTENTS

FOS—40 Litho in U.S.A.

FOS—40 Litho in U.S.A.

POWER TRAINS—How They Work / CHAPTER 1

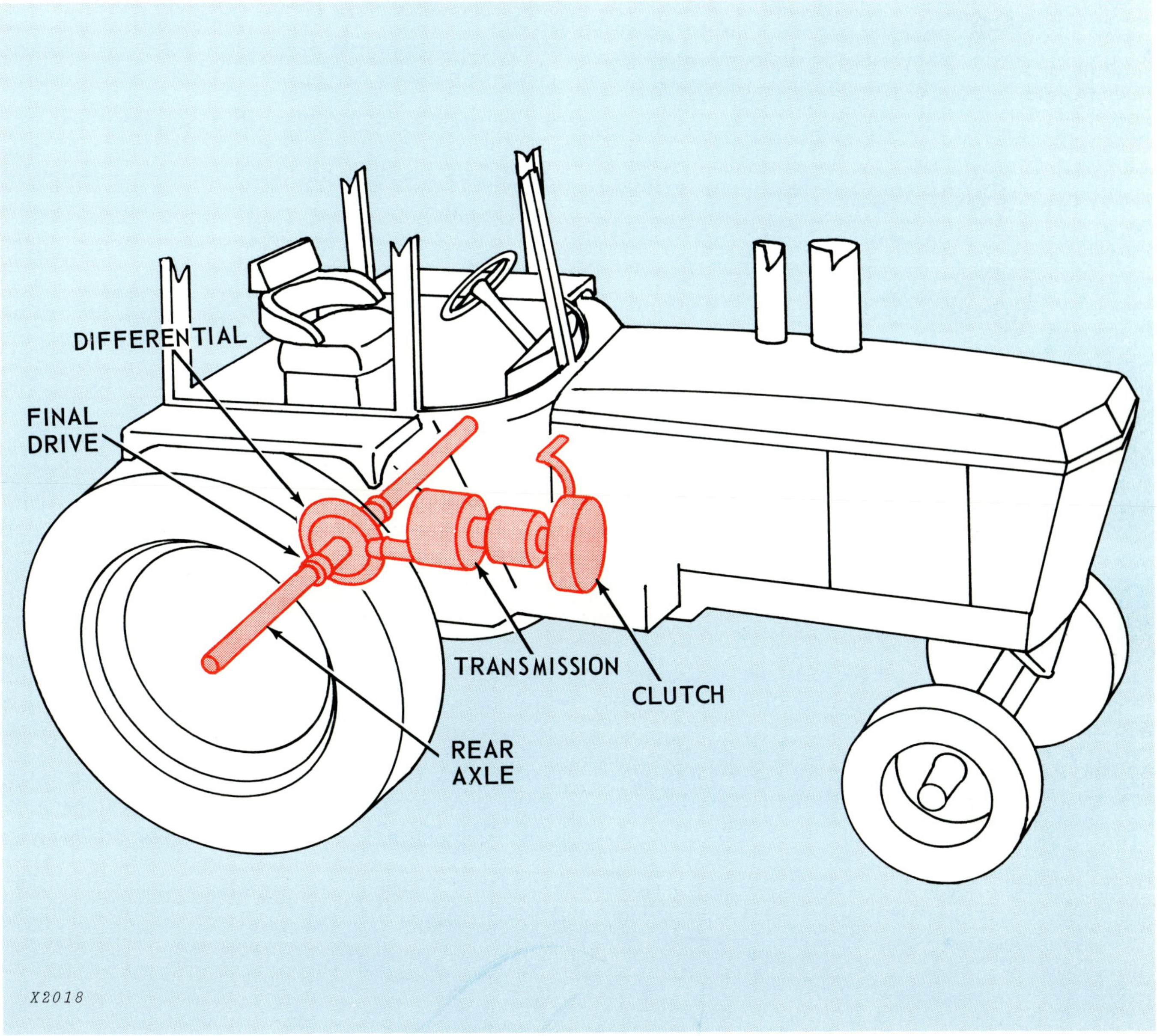

Fig. 1—Power Train For a Tractor

INTRODUCTION

Engine power is transmitted to the drive wheels or output shaft of a machine by the **power train.** The power train does four jobs:

1) Connects and disconnects power

2) Selects speed ratios

3) Provides a means of reversing

4) Equalizes power to the drive wheels for turning

To do these jobs, five components are needed:

- **Clutch—to connect and disconnect power**
- **Transmission—to select speeds and direction**
- **Differential—to equalize power for turning**
- **Final Drives—to reduce speed and increase torque to axle**
- **Drive Wheels—to propel the machine**

Some components are also needed to help drive the machine, but these are the basic power train components.

Now let's follow the flow of power through a simple power train.

HOW A POWER TRAIN WORKS

The first part of a power train is the **clutch.**

Its job is to disconnect the engine from the power train, allowing the engine to run while the machine is standing still. The clutch also engages this power to start up the machine.

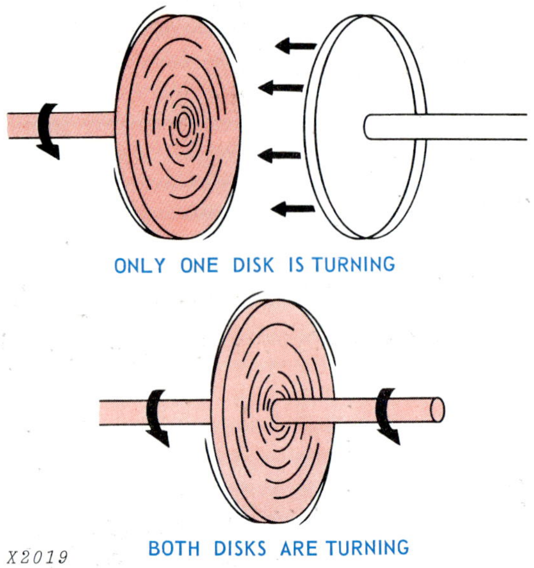

Fig. 2—How The Clutch Works

Think of a clutch as two disks, each on a shaft (Fig. 2).

As long as the disks are not touching, we can spin one as fast as we want to without affecting the other.

But if we move them together when one is spinning, the other will begin to turn and then both shafts will turn as one unit.

This is the principle of the disk clutch as used on many machines.

In an actual disk clutch (Fig. 3), the disks are forced together by strong springs, and are separated by pushing down on the clutch pedal.

Some machines have automatic drives in which the clutch is eliminated. We will discuss these later.

The **transmission** allows us to change the speed of the drive wheels in relation to the engine.

This allows slower speeds when starting up the machine, and faster speeds to match various loads and conditions.

In automobiles, we need more turning force on the rear wheels to start moving than we need for cruising along a good highway. The transmission gives us this increased turning force and also allows the engine to accelerate. The latter is important because the engine does not develop very much power at low rpms. When the car has picked up speed, the transmission is shifted to change the speed ratio between the engine and the wheels, and eventually is shifted into direct drive.

In tractors and other field machines, there is an even greater speed reduction to give more turning force for traction and for pulling loads.

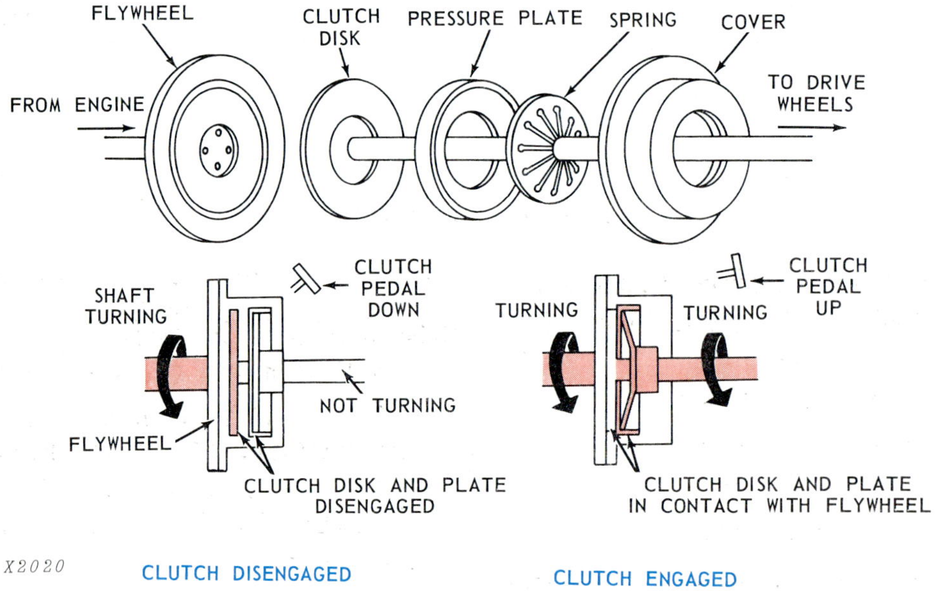

Fig. 3—Disk Clutch In Operation

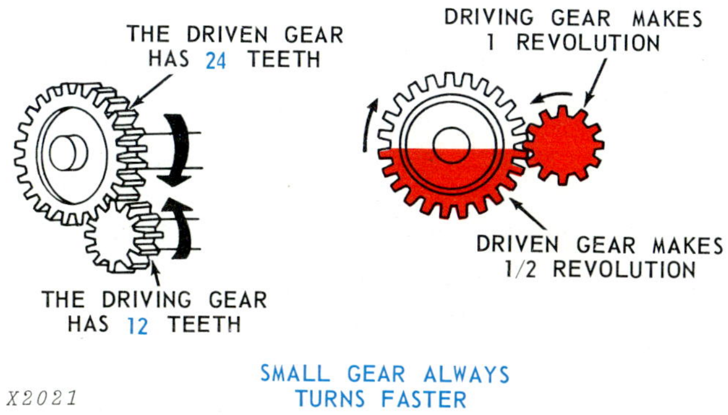

Fig. 4—How The Transmission Selects A Speed

The transmission is a system of **gears.**

Suppose we have a small gear with 12 teeth driving a larger gear with 24 teeth (Fig. 4).

When the first gear has made one complete revolution, it has gone around the equivalent of 12 teeth. The second one has gone around the same distance—12 teeth—but this means only *one-half* a revolution for the larger gear.

As a result, the second gear and its shaft always turn at one-half the speed of the first gear and its shaft.

The point to remember is that **smaller gears always turn at a faster rate.**

This is the principle of the transmission—several combinations of gears are arranged so that we can select the speed we want to use at any moment.

For **low or first gear,** a small gear on the input shaft drives a large gear on another shaft (Fig. 5).

This reduces the speed and increases the turning force.

Then a small gear on the second shaft drives a large gear on the drive shaft which goes to the driving axle.

This reduces the speed and increases the turning still more, giving a higher gear ratio for starting up or heavy pulling.

For **second gear,** we can use the same first pair of gears as in low.

However, we disconnect the second pair of gears and drive through two other gears (Fig. 6).

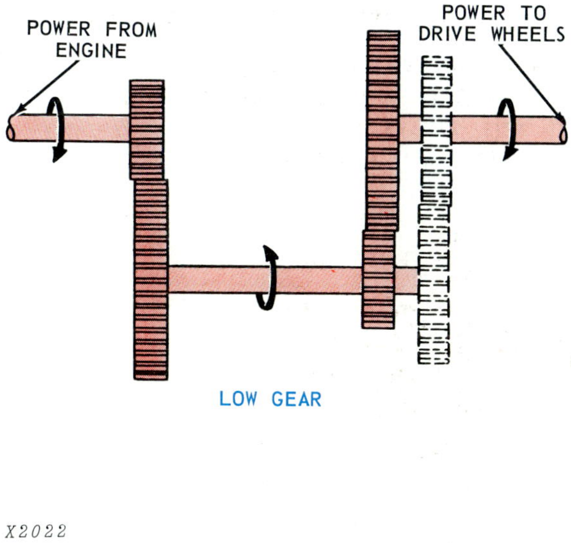

Fig. 5—Transmission In Low or First Gear

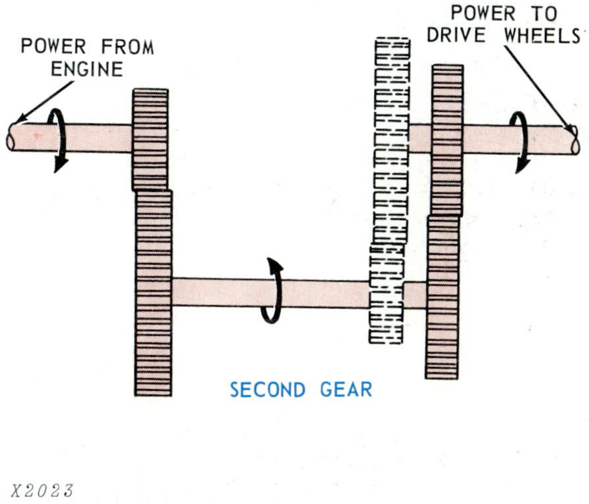

Fig. 6—Transmission In Second Gear

These gears are arranged so that the larger one drives the smaller, so there is less overall speed reduction than in first gear.

For higher gears, the gear ratio is cut further by using other gear combinations. In fact, high gear for an automobile is normally a direct drive with no reduction (a 1 to 1 ratio).

 Litho in U.S.A.

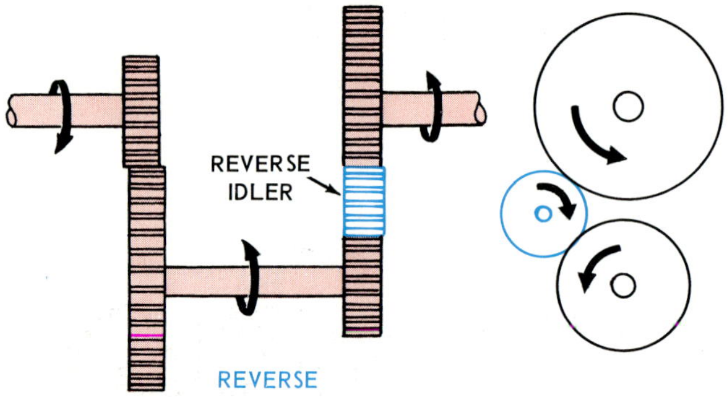

Fig. 7—Transmission in Reverse Gear

Reverse gear is very much like first, giving about the same ratio and using the same four gears. However, it also uses an extra gear called a reverse idler (Fig. 7) which causes the drive shaft to turn in the opposite direction.

All the gears are mounted in a metal case filled with oil to lubricate the gears and bearings.

The various speeds are selected by moving a shift lever in the driver's compartment.

Other types of transmissions, covered later in this manual, are automatic drives which use hydraulic power to aid in controlling the power train.

Now let's complete the power train from the transmission to the drive wheels.

In the rear axle we have two sets of gears. The first—the **ring gear and pinion**—simply transmits the power around the corner to the axles as shown in Fig. 8.

This allows the propeller shaft to drive the axles which are at right angles to it.

If we never turned a corner, that is all the gearing we would need at the rear axle. But when we turn a corner, the outside wheel has to travel farther than the inside wheel and has to turn faster.

The **differential** (Fig. 9) allows each wheel to travel at a different speed and still propel its own load.

When the machine is moving straight ahead, *both wheels are free to rotate* as shown at left.

Engine power comes in on the pinion gear and rotates the ring gear. The four bevel pinions and

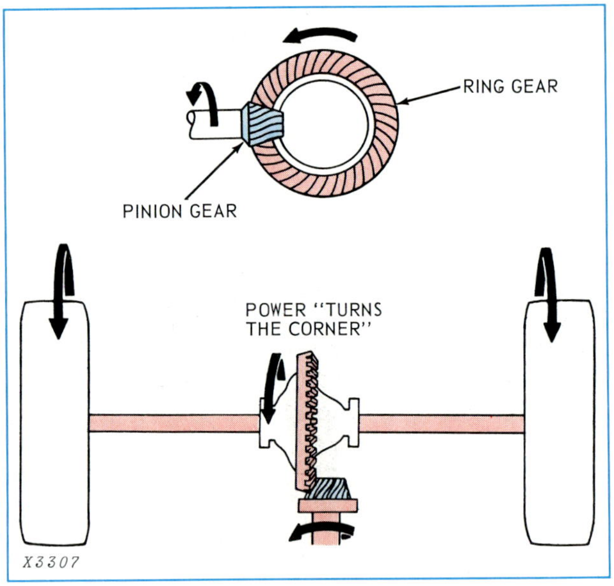

Fig. 8—Ring Gear and Pinion For Rear Axle

the two bevel gears are carried around by the ring gear and all gears rotate as one unit. Each axle receives the same rotation and so each wheel turns at the same speed.

When the machine turns a sharp corner, only *one wheel is free to rotate* as shown at right in Fig. 9.

Again, engine power comes in on the pinion gear and rotates the ring gear, carrying the bevel pinions around with it. However, the right-hand axle is held stationary and so the bevel pinions are forced to rotate on their own axes and "walk around" the right-hand bevel gear.

Since the bevel pinions are in constant mesh with both bevel gears, the left-hand bevel gear is forced to rotate because it is subjected to the turning force of the ring gear which is transmitted through the bevel pinions.

During one revolution of the ring gear, the left-hand bevel gear makes *two* revolutions—one with the ring gear and another as the bevel pinions "walk around" the other bevel gear.

As a result, when the drive wheels have unequal resistance applied to them, the wheel with the least resistance turns more revolutions. As one wheel turns faster, the other turns slower by the same amount.

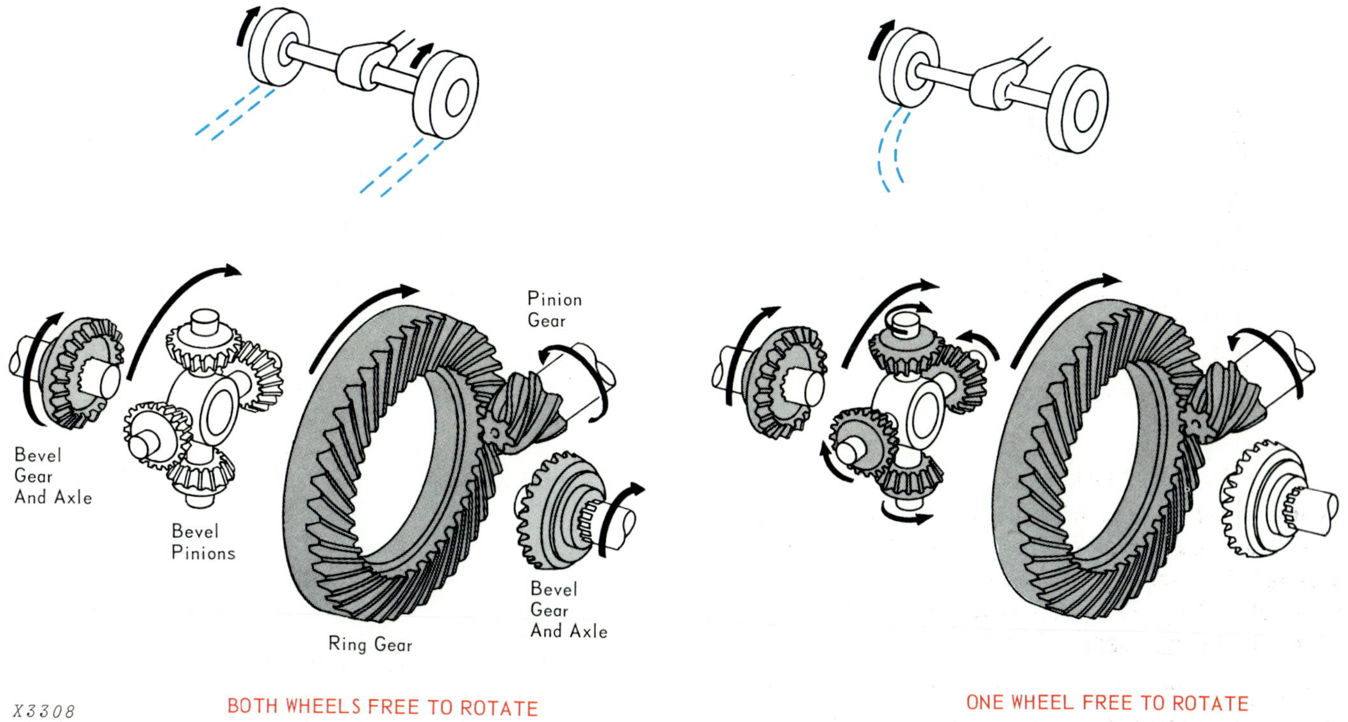

Fig. 9—Differential

Differential action can be a disadvantage when a farm tractor is plowing, for example. One wheel may lose resistance on slick ground and start spinning. To prevent a loss of power, a differential **lock** is often used to lock the two axles together until the slippery spot is passed, when they are released.

Some machines have **final drives** on the axles to further reduce speed and increase torque. They are most common on large machines which handle very big loads. Bull gears and planetary gears are the main types of gearing.

The **axles,** of course, rotate the wheels to make the machine move, which completes our basic power train.

A complete power train in a modern tractor is shown in Fig. 10. The clutch is shown at the right, the transmission in the center, and the differential at the left.

For details on actual power trains, see these chapters:

Chapter 2—Clutches

Chapters 3 and 4—Transmissions

Chapter 7—Differentials

Chapter 8—Final Drives

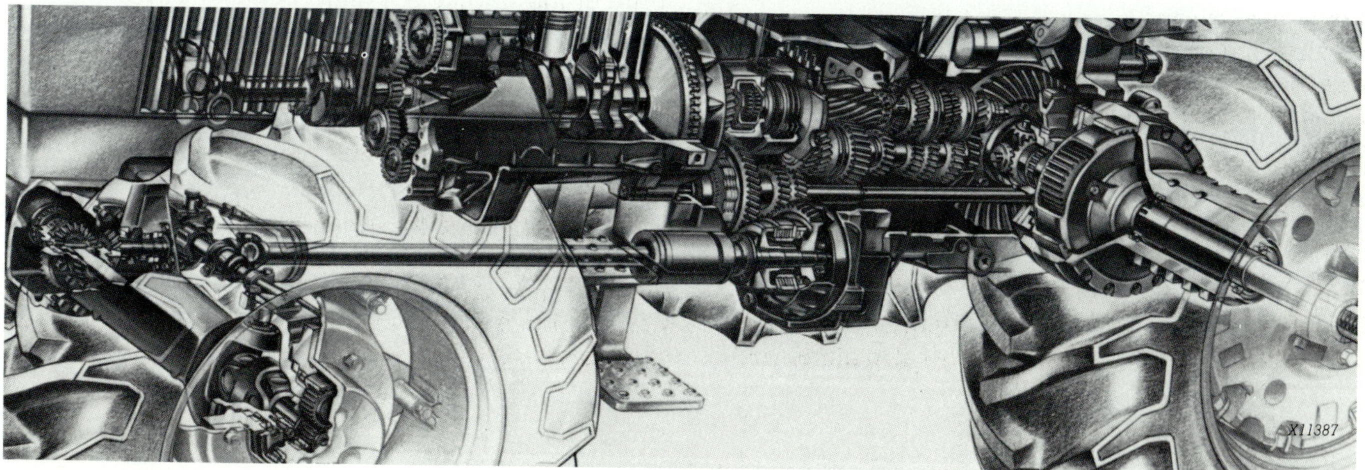

Fig. 10—Tractor Power Train

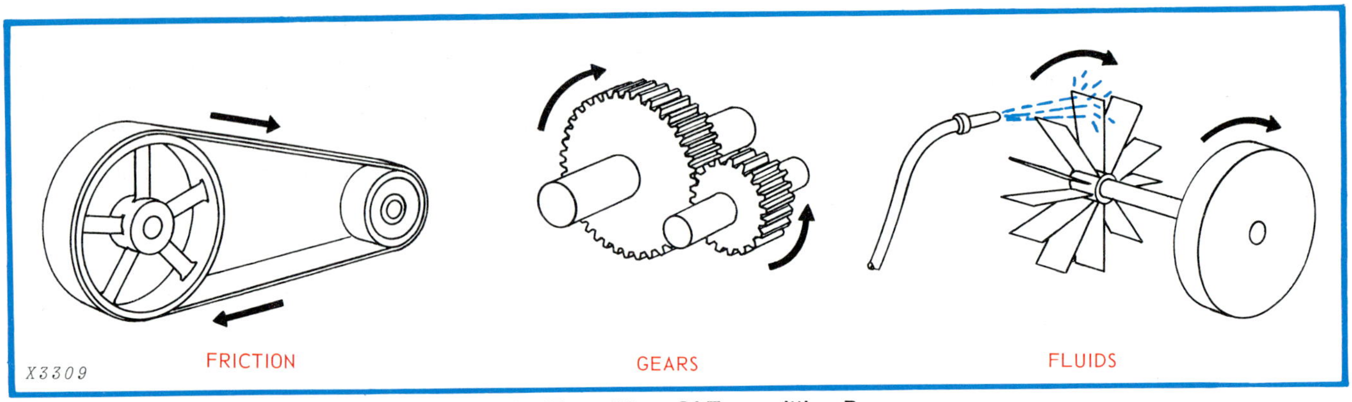

Fig. 11—Three Ways Of Transmitting Power

HOW POWER IS TRANSMITTED

We have seen how power is transmitted in a basic power train.

The clutch uses *friction* to transmit power, while the transmission uses *gears.* A third way to transmit power is by *fluids,* as in automatic drives.

So the three basic ways of transmitting power are:

- **Friction** *(Wheels and Belts)*
- **Gears** *(In Mesh)*
- **Fluids** *(Water Wheel Principle)*

Let's take a look at each type.

FRICTION DRIVES

Friction is caused by bringing together two surfaces made from materials that will transmit motion from one to the other. Slippage between the surfaces is a matter of engineering design. For our purposes, let's assume no slippage takes place.

Friction devices include wheels and belts.

Wheel Friction

A simple form of useable friction is obtained by a rubber-tired wheel contacting the ground (Fig. 12, top). If you rotate the wheel by hand, you will feel the rotary force tend to move you forward. The friction between the tire and the ground is transmitting your force to the ground. Since you can't move the ground, the wheel and you must move.

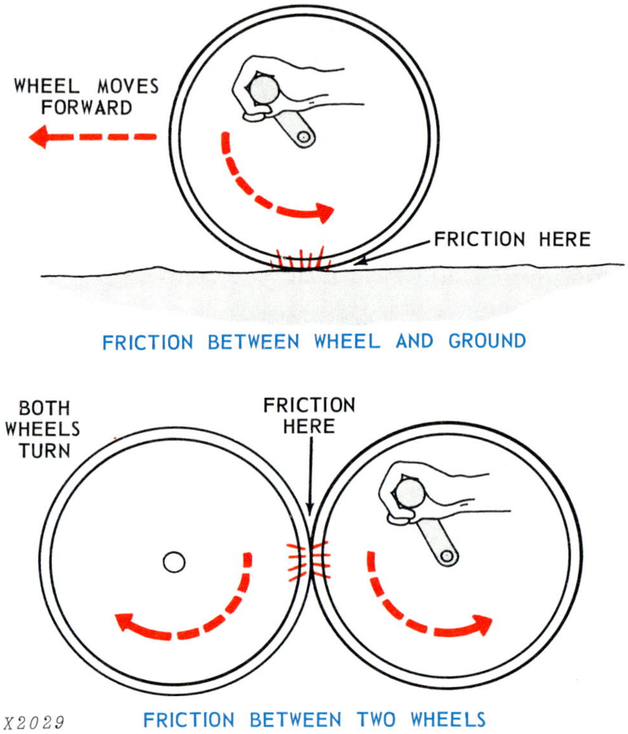

Fig. 12—Wheel Friction

Now suppose we suspend the first wheel above ground so it is free to rotate. Let's also suspend a second wheel in contact with the first (Fig. 12, bottom). Now if you apply a rotating force to the first wheel, the second one will rotate. Observe that the two wheels rotate in opposite directions. Thus, we can transmit power from one point to another by means of friction.

The engine clutch is a prime example of transmitting power by friction.

 Litho in U.S.A.

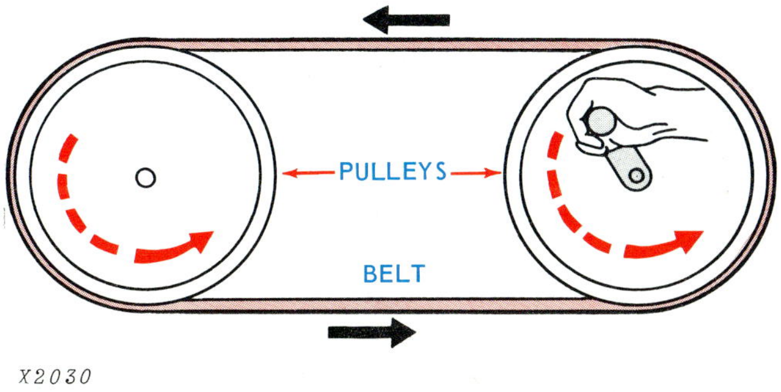

Fig. 13—Belt Friction

Belt Friction

Another friction drive also has two wheels, called sheaves or *pulleys.* This arrangement permits the axles to be separated by some distance. The friction coupling is provided by the continuous loop of the *belt* as shown in Fig. 13.

This gives a more efficient transmission of power than two wheels because of more surface contact. When one pulley is turned, the belt transmits the power by friction to the other wheel as shown.

Belt drives are covered in detail in Chapter 10.

GEAR DRIVES

Gears are the most common way to transmit power.

When two gears are in mesh, all slippage is gone. For this reason, gears are widely used for high-power applications.

The chain drive is another variation on the gear drive. The gears (sprockets) are not in mesh but are connected by a linked chain which also eliminates slippage (see Chapter 10 for details).

Gears will be covered in detail in the next part of this chapter.

FLUID DRIVES

Fluid drives are both the oldest and most modern ways of transmitting power.

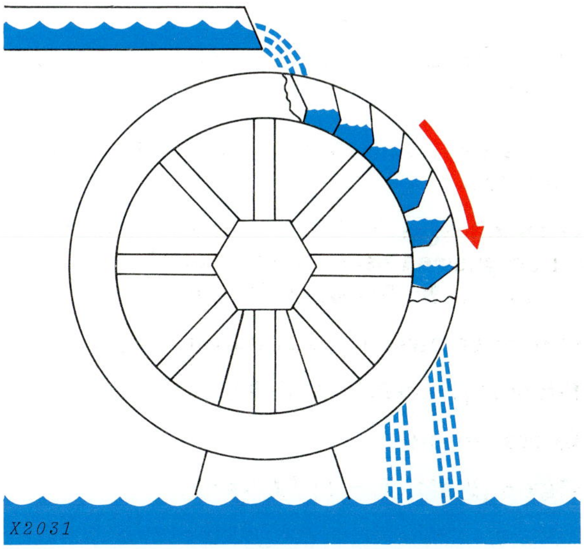

Fig. 14—Fluid Drive In Water Wheel

The old mill water wheel (Fig. 14) is turned by water falling into its buckets.

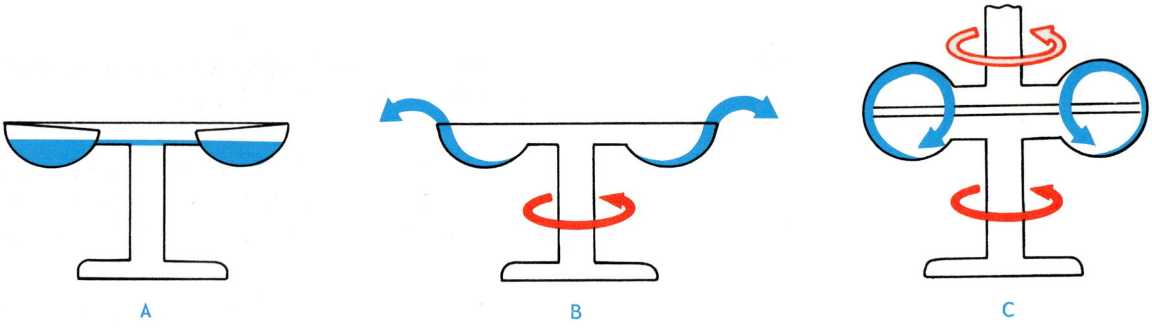

Fig. 15—Fluid Coupling Between Two Shafts

In modern automatic drives, fluid is used as a coupling for power (Fig. 15).

In A, we start with a disk which contains vanes. If we supply a fluid, the vanes fill as shown.

In B, we rotate the disk at high speed and centrifugal force causes the fluid to move out as shown.

In C, we place a second disk above and near the first disk. Fluid now flows into the second disk as shown. The force of the fluid on the second disk will cause it to rotate in the same direction as the first one, forming a fluid coupling which transmits power.

When the disks are sealed and the fluid is under pressure, a solid coupling is formed.

This is the basic principle of fluid drives such as the fluid clutch and torque converter.

In Chapter 6, we will see how these parts are modified to *increase* the torque in a torque converter.

This completes our look at the basic power transmitting devices—**friction, gears,** and **fluids.**

Because gears are so basic to power trains, let's take a detailed look at them now.

GEARS

The basic elements of almost all conventional power trains are **gears.**

Gears are simply a means of applying twisting force—or **torque**—to rotating parts.

The amount of torque you can get from a source of power is proportional to the distance from the center at which it is applied.

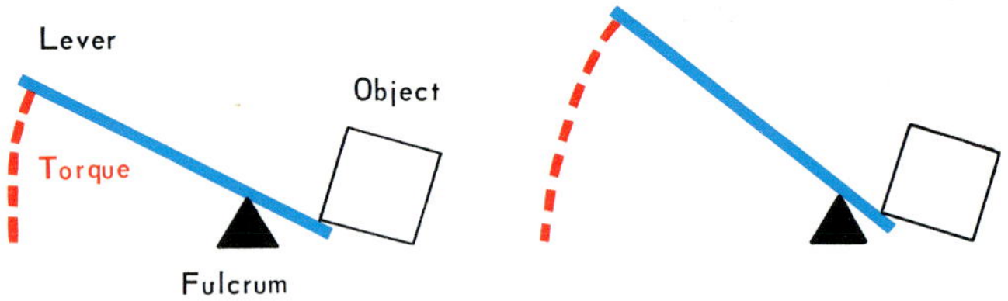

Fig. 16—How Torque Varies With Levers

In Fig. 16, the lever has more torque as the fulcrum gets closer to the object of power application (at right). But the lever must also be rotated farther to get this torque.

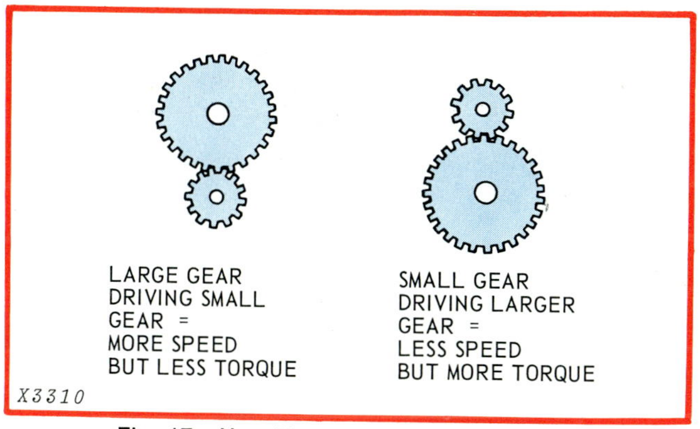

Fig. 17—How Torque Varies With Gears

The same principle is used for gears in mesh (Fig. 17).

A small gear will drive a large gear more slowly but with greater torque.

The gears in a transmission are selected to give the operator a choice of both *speed* and *torque.*

Lower gear range = less speed but more torque.

Higher gear range = less torque but more speed.

The final choice of gears is a compromise to give enough of each to suit the needs of the machine and its operator.

GEAR RATIOS

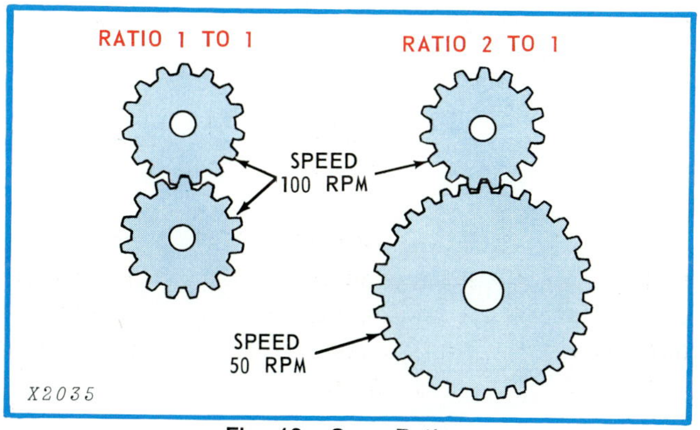

Fig. 18—Gear Ratios

Gear ratio is a measure of the changes in speed and torque in the gear train.

The gear ratio between the gears in Fig. 18 is the ratio of the number of teeth on the lower gear to the number of teeth on the upper gear.

TYPES OF GEARS

A variety of gear types are used to meet the demands of speed and torque.

Gears are normally used to **transmit torque from one shaft to another.** These shafts may operate in line, parallel, or at an angle to each other.

Meshing gears must also have teeth of the same size and design. And at least one pair of teeth must be engaged at all times. Some tooth designs allow for contact between more than one pair of teeth.

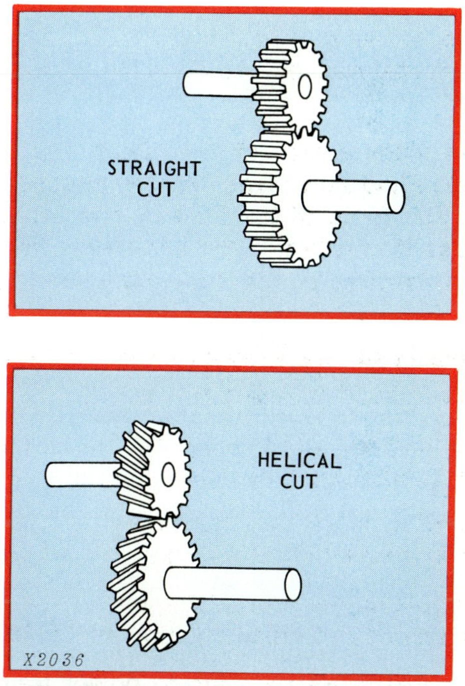

Fig. 19—Gear Teeth—Two Major Types

Gears are normally classified by:

1) Type of teeth.

2) Surface on which teeth are cut.

Fig. 19 shows the two major types of gear teeth—**straight cut** and **helical cut.**

 Litho in U.S.A.

The most common types of gears are shown in Fig. 20.

STRAIGHT SPUR GEARS—These gears have straight teeth cut parallel to the axis of rotation. Normally, mating gears have one to two pairs of teeth engaged at all times. These gears are noisier and so are used mainly for slow speeds to avoid excessive vibration.

Uses: Straight spur gears are used in simple devices such as hand or powered winches. Sliding gear transmissions also use them for speed changes because they can be easily shifted by sliding on the shaft from one gear to another.

HELICAL SPUR GEARS—The teeth of a helical spur gear are cut obliquely across the perimeter of the gear instead of straight. Engagement between two teeth starts at the tooth tip of one gear and rolls down the teeth to the trailing edge. This angular contact tends to cause side thrusts which the bearings must absorb. However, helical spur gears are quieter in operation and have greater strength and durability than straight spur gears—because the contacting teeth are longer.

Uses: Helical spur gears are widely used in machine transmissions today because they are quieter at high speeds, and are durable.

HERRINGBONE GEARS—Herringbone gears are really double helical gears with teeth angles reversed on opposite sides. This causes the thrust produced by one side to be counterbalanced by the thrust produced by the other side. The two sets of teeth are often separated at the center by a narrow gap for better alignment and to prevent oil from being trapped at the apex.

Uses: Herringbone gears are best suited for quiet, high-speed, low-thrust applications where heavy loads are applied. Large turbines and generators frequently use herringbone gears because of their durability.

PLAIN BEVEL GEARS—These gears permit the power flow to "turn a corner". The gear teeth are cut straight on a line with the shaft but at some angle between perpendicular and parallel to the shaft. The two gears are commonly called the "ring gear" (larger driven gear) and "pinion gear" (smaller driving gear).

Uses: Like the straight spur gears, the plain bevel type is used in slow-speed applications which are not subject to high impact forces. Handwheel controls which must operate some remote device at an angle use straight bevel gears.

SPIRAL BEVEL GEARS—These gears were developed for use where higher speed and strength were required while changing the angle of power flow. Their teeth are cut obliquely on the angular faces of the gears. The angle is determined by the angle between the two shafts.

Uses: Farm and industrial machines use these gears in ring gear-and-pinion sets at the drive axles. They not only change the angle of power flow, but they also reduce speed and increase force.

HYPOID GEARS—The hypoid gear resembles the spiral bevel gear but the pinion drive gear (smaller) is located below the center of the ring gear (larger). Teeth and general construction are the same as the spiral bevel gear.

Uses: The most common use for hypoid gears is in modern automobile differentials. Here they allow for lower body styles by lowering the transmission drive shaft.

PLANETARY GEARS—These are gear sets in which an outer ring gear has internal teeth which mate with teeth on smaller planet gears. These gears in turn mate with a center or sun gear. Many changes in speed and torque are possible, depending on which parts are braked and which are driven. (Operation of planetaries is explained on page 1-12.)

Uses: Planetary gears are widely used in transmissions because each set is capable of more than one speed change. The gear load is spread over several gears, reducing stress and wear on any one gear. Final drives in heavy machinery can also use these gears.

WORM GEARS—The worm gear is actually a screw (inclined plane). It is capable of high speed reductions in a compact space. The mating gear has teeth which are curved at the tips to permit a greater contact area. Power is supplied to the worm gear, which drives the mating gear. Worm gears usually provide right-angle power flows.

Uses: The most common use for the worm gear is where the power source operates at high speed and the output is slow speed with high torque. Many steering mechanisms use a worm gear connected to the streering shaft and wheel and a partial (sector) gear connected to the steering linkage. Small power hand tools frequently use a high-speed motor with a worm gear drive.

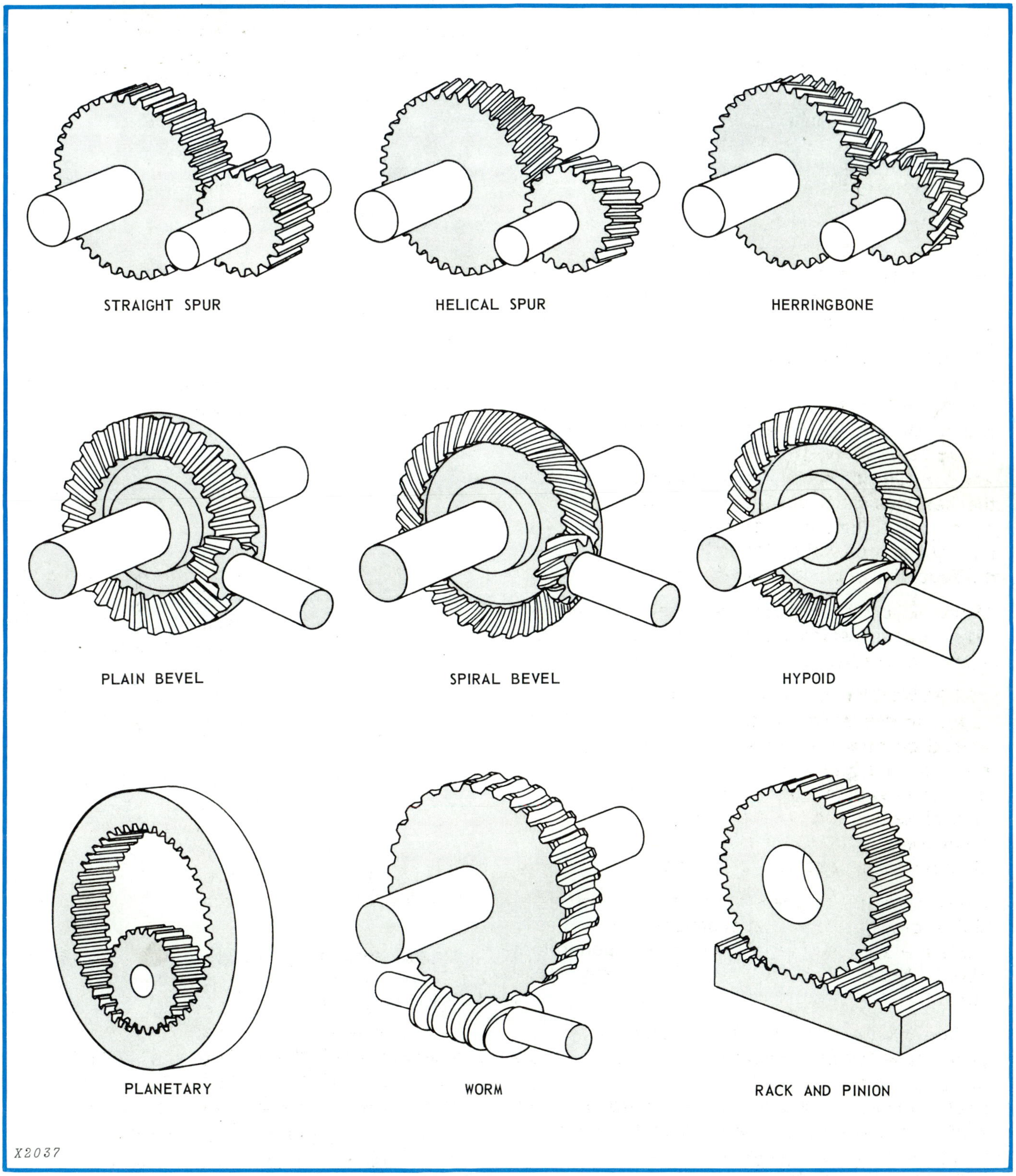

Fig. 20—Types of Gears

RACK AND PINION GEARS—This set of gears converts straight-line motion into rotary motion and vice versa. Rack and pinion gears also change the angle of power flow with some degree of speed change. The teeth on the rack are cut straight while those on the pinion are curved.

Uses: The rack and pinion can provide control of arbor presses and other devices where slow speed is involved.

PLANETARY GEARS—HOW THEY WORK

Planetary gears are simple in design, but their operation is not always well understood.

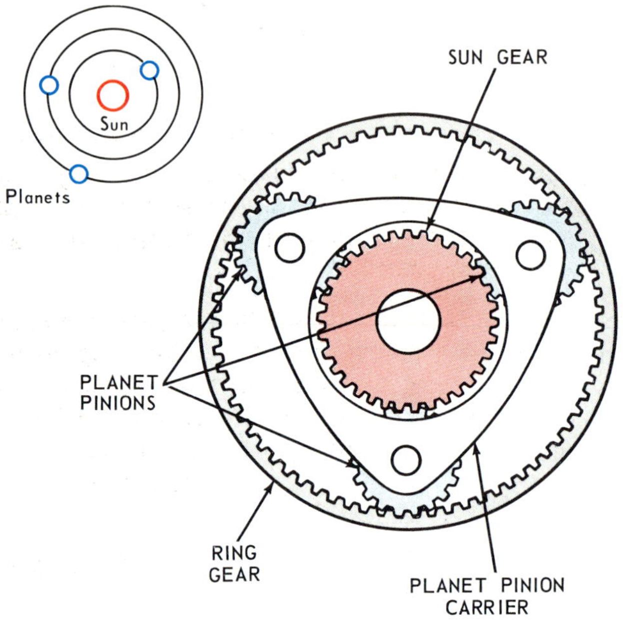

X2038

Fig. 21—Simple Planetary System

Planetary gears are similar to our solar system (Fig. 21). The **planet pinion gears** each turn on their own axes while rotating around the **sun gear.** This is much like the earth and other planets rotating around the sun. The pinion gears in turn mesh with the inside of the **ring gear.**

Notice that the sun gear, the planet pinions, and the ring gear are constantly in mesh.

The planet pinions are mounted on shafts in the carrier, and can rotate on their own axes to "walk around" the sun gear or the ring gear.

When power is applied to drive the sun gear or the planet pinion carrier, the entire system will rotate as a unit unless a restraining force is applied to hold one of the other two members of the system stationary.

When power is applied to one member of the planetary system, and a brake is applied to restrain a second member from turning, the remaining part will become a power output source as illustrated by the following examples.

When The Sun Gear Is Driven

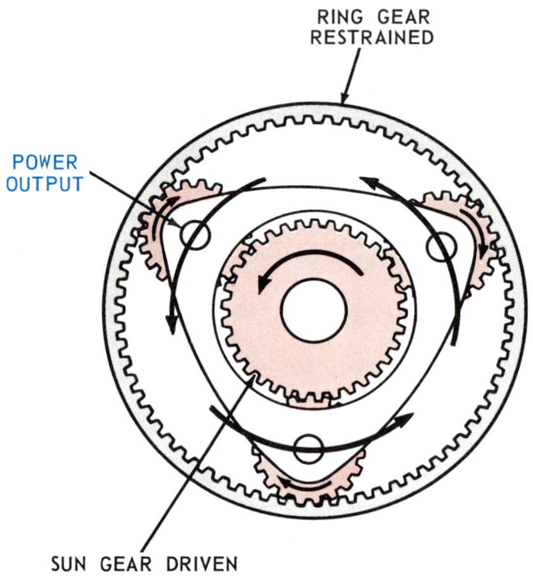

X2039

Fig. 22—Power Flow in Planetary When Sun Gear is Driven

When the sun gear is driven (Fig. 22) and a brake is applied to the ring gear, the planet pinions "walk around" the ring gear, forcing the planet pinion carrier to rotate in the same direction as the sun gear, but at a slower speed.

When Planet Pinion Carrier Is Driven

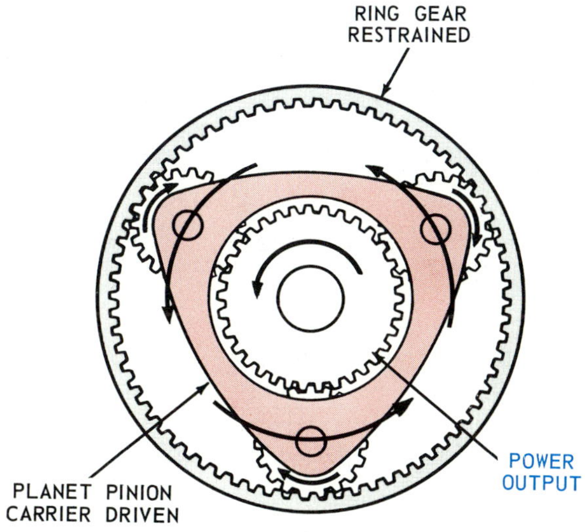

Fig. 23—Power Flow In Planetary When Planet Pinion Carrier Is Driven

When the planet pinion carrier is driven (Fig. 23) and a brake is applied to the ring gear, the planet pinions "walk around" the ring gear, forcing the sun gear to rotate in the same direction at a higher speed.

In both cases above, one member is driven, one is braked, and the third member becomes the power output.

Depending upon which member is driven and which is braked, the planetary gives different speeds.

Double Planet Pinion Set

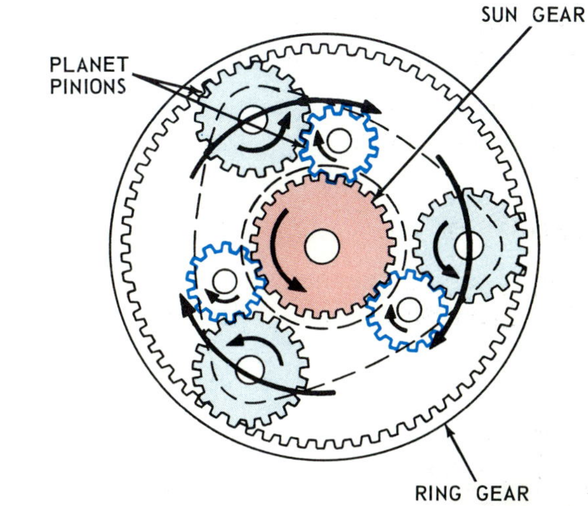

Fig. 24—Double Planet Pinion Set Giving Reverse Speed

When a second set of planet pinions (Fig. 24) is added to the simple planetary system, so that the two sets of planet pinions are in mesh, reverse speeds can be obtained.

Now when power is applied to the planet pinion carrier and a brake is applied to the ring gear, the planet pinions in mesh with the ring gear are forced to rotate on their axes, driving the inner planet pinions which in turn force the sun gear to rotate in the reverse direction from the planet pinion carrier.

The planetary system we have described will give low, high, and reverse speed ranges.

For an example of how this works in a planetary transmission, see Chapter 4.

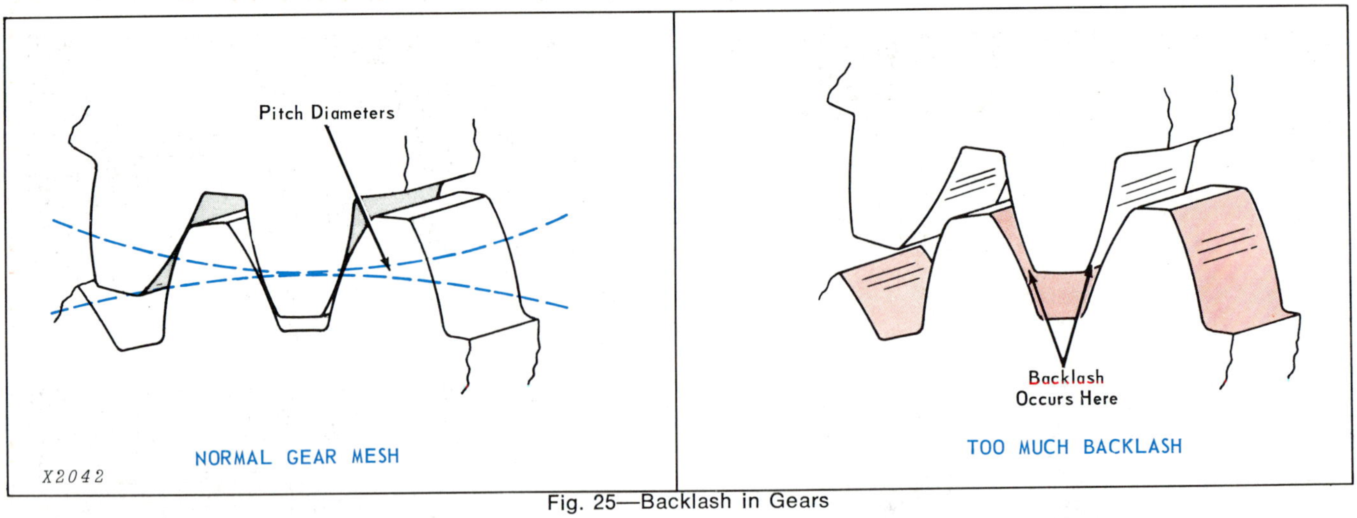

Fig. 25—Backlash in Gears

BACKLASH IN GEARS

Backlash is the clearance or "play" between two gears in mesh.

Too much backlash can be caused by worn gear teeth, an improper meshing of teeth, or bearings which do not support the gears properly.

Too much backlash can result in severe impact on the gear teeth from sudden stops or reverses of the gears. Broken gear teeth and gears bouncing under impact also results from too much backlash.

Too little backlash causes excessive overload wear on gear teeth. This could result in premature gear failure.

Fig. 25 shows the normal gear mesh and gear mesh which permits too much backlash.

On the *normal gears* in Fig. 25, the clearance of the teeth at the pitch diameters is very small.

However, on the worn gears with *too much backlash,* forces cause a greater movement and a higher impact which can break the teeth or at least cause the gears to bounce.

Backlash, endplay and preload on gears and shafts is covered later in this chapter under "Adjusting The Gear Train".

GEAR WEAR

New gear teeth have slight imperfections, but they normally disappear during break-in as the teeth are oiled and polished. After that, the teeth should have a long service life.

However, when lack of lubrication or other factors cause a gear to fail, we can examine the failure and determine the cause.

The major types of gear tooth wear and failures are shown in Fig. 26.

NORMAL WEAR—This is the normal polishing of the gear teeth as they operate. The polished surface should extend the full length of the tooth from near the root (or bottom) to the tip of the tooth. Gears which are manufactured properly, well lubricated, and not overloaded or improperly installed will show this condition after many hours of servoce.

ABRASIVE WEAR—Surface injury caused by fine particles carried in the lubricant or embedded in the tooth surfaces. The causes are metal particles from gear teeth, abrasives left in the gear case, or sand and scale from castings.

SCRATCHING—This is often found on gears which handle heavy loads at slow speeds. It is caused by particles of metal flaking off the gears which are larger than the abrasive particles. Generally, it indicates the wrong gear design for the load. (Do not confuse this with scoring.)

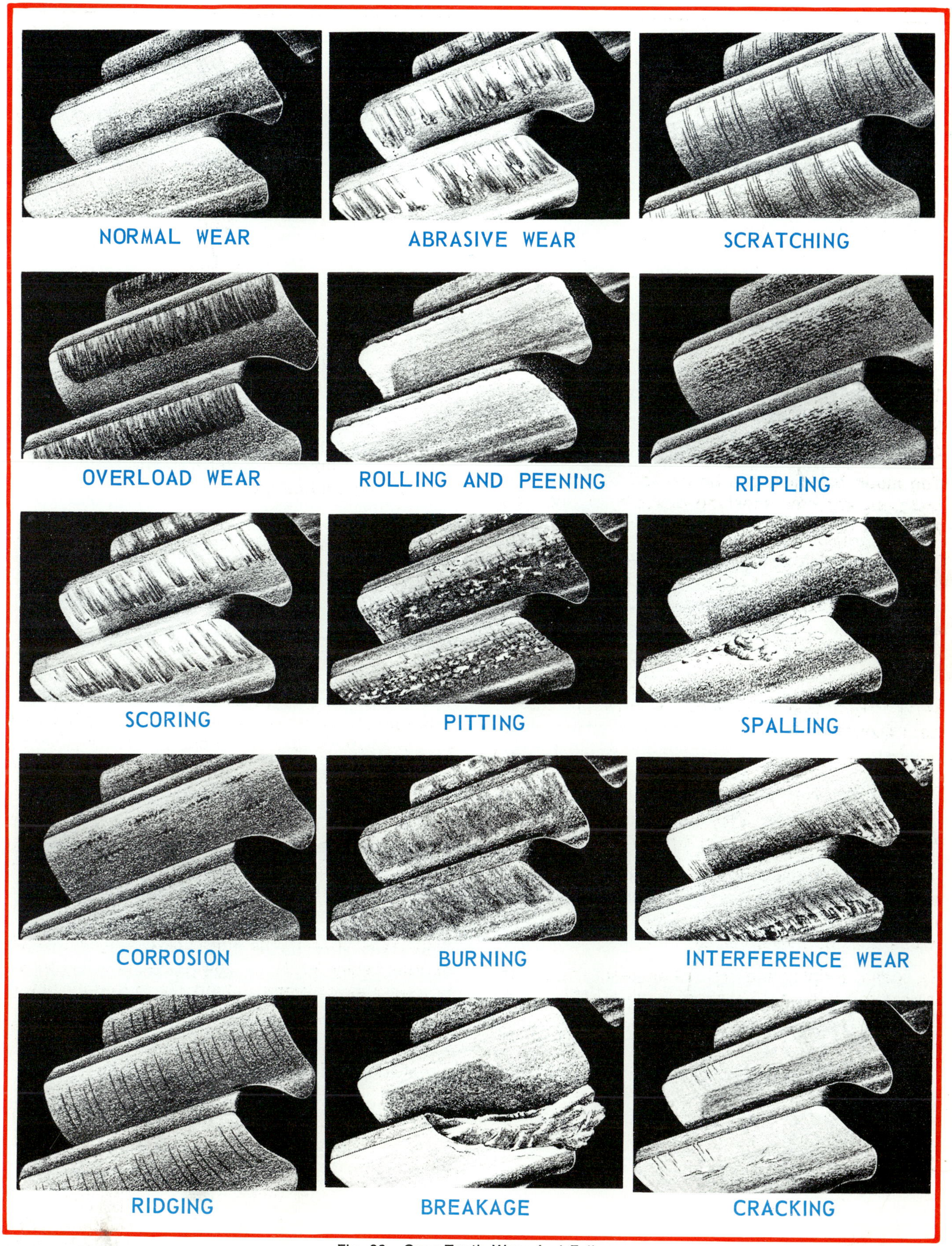

Fig. 26—Gear Tooth Wear And Failures

OVERLOAD WEAR—If the contact surface is worn but smooth, the gears have been overloaded and metal has been removed by the sliding pressure causing a depression in the length of the teeth. Continuous use will result in backlash and severe peening which may be misleading as to the real cause of the wear.

ROLLING AND PEENING—*Rolling* is the result of overload and sliding which leaves a burr on the tooth edge. Too little bearing support or too ductile a metal results in plastic flow of the metal due to sliding pressure. *Peening* is the result of backlash and the force causing a tooth to hammer on another with tremendous impact. In these cases, lubricants are forced out and metal bears directly on metal.

RIPPLING—This is a wavy surface or "fish scales" on the teeth at right angles to the direction of slide. It may be caused by surface yielding due to "slipstick" friction resulting from lack of lubrication, heavy loads, or vibrations.

SCORING—This is caused by temperature rise and thinning or rupture of the lubricant film as from too heavy loads. Pressure and sliding action heats the gear and permits metal transfer from one tooth to the face of another. As the process continues, chunks of metal loosen and gouge the teeth in the direction of the sliding motion. The temperature rise here is slow and not as high as *burning* wear (below).

PITTING—Gear teeth should not show pitting. Very minute or micro-pitting can occur and would appear as a gray surface which may advance slowly to an actual pitted condition. This type of condition is sometimes associated with thin oil film, possibly due to high oil temperatures.

SPALLING — Spalling is a common wear condition which starts with fine surface cracks and eventually results in large flakes or chips leaving the tooth face. Improperly case-hardened teeth are most often subject to this kind of damage due to the brittle nature of the metal. Spalling may occur on one or two teeth but the chips may cause other damage to the remaining teeth.

CORROSION—Corrosive wear results in an erosion of the tooth surfaces by acid. The acid is formed by moisture combining with lubricant impurities and atmospheric contaminants. Generally, the surfaces become pitted, causing an uneven surface and distribution of stresses which lead to chipping and spalling.

BURNING—Burning is usually caused by the complete failure of lubricants or a lack of lubrication. During high stress and sliding motion, friction develops rapid heating and the temperature limits of the metal are exceeded. Burned gear teeth are extremely brittle and easily broken.

INTERFERENCE WEAR—This type of wear can be caused by misalignment of gears which places heavy contact on small areas. Also, mating of two gears with teeth not designed to work together will cause interference wear. More than one wear pattern may show up, as at teeth tips and roots.

RIDGING—These are scratches appearing near one end of a tooth, especially on a hypoid pinion gear. This can be caused by excessive loads or lack of lubrication, or by gear not properly heat treated during manufacture.

BREAKAGE—Broken teeth may be the result of many defects. Make a close study of the other teeth before judging the cause. Breakage can be caused by high impact forces or defective manufacture. To determine if breakage is due to overload or fatigue, examine the broken area closely. If the break shows fresh metal all over the break, an impact overload was the cause. If the break shows an area in the center of fresh metal with the edges dark and old-looking, the breakage was due to fatigue which started with a fine surface crack.

CRACKING—These failures tend to be caused by improper heat treating during manufacture. Improperly machined tooth root dimensions can also result in cracking. Most heat treat cracks are extremely fine and do not show up until a gear has been used for some time.

SUMMARY: GEARS

Gears are very basic to the function of power trains. In general, excessive backlash or endplay, or improper lubrication can lead to gear failures in the system.

Always be aware of the forces which are at work when gears are operating under load.

Alignment and fit are very important to insure that each tooth absorbs its share of the load and that the gears are supported to resist thrust and twisting forces.

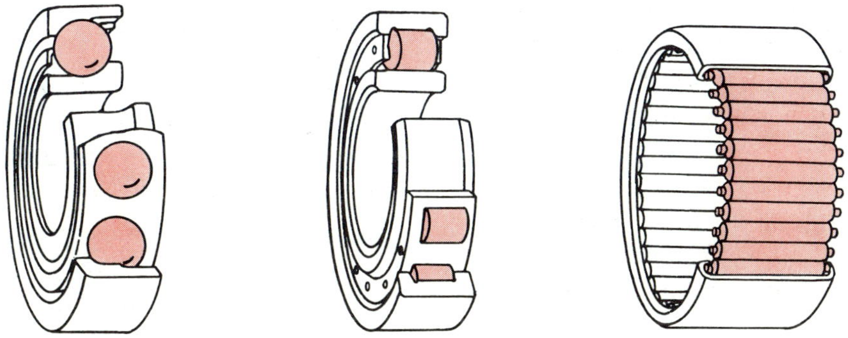

Fig. 27—Bearings—Three Types

BEARINGS

Bearings have two major jobs in a power train:

1) Reduce friction

2) Support a rotating shaft

We are mainly concerned with *anti-friction* bearings, those that give a rolling contact between mating surfaces (Fig. 27). There are three main types:

- **Ball**
- **Roller**
- **Needle**

All these bearings are made of:

1) Two hardened-steel rings called races (Fig. 28).

2) Balls, rollers or needles which roll between the two races.

3) Optional separators to space the rolling elements around the bearing.

On some bearings, the outer or inner race is omitted. Then the rolling elements are in direct contact with the shaft or other mounting (as in most needle bearings).

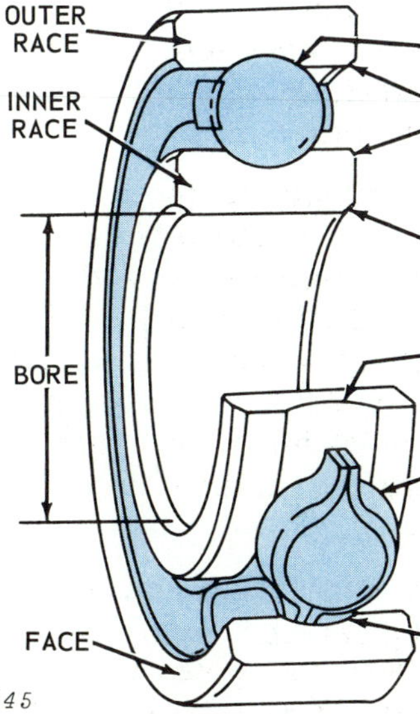

Fig. 28—Basic Parts Of A Bearing

When two races are used, one race is normally pressed or fixed on a shaft or in a bore, while the other race is free to turn with the rolling elements. This is part of the "anti-friction" feature of these bearings.

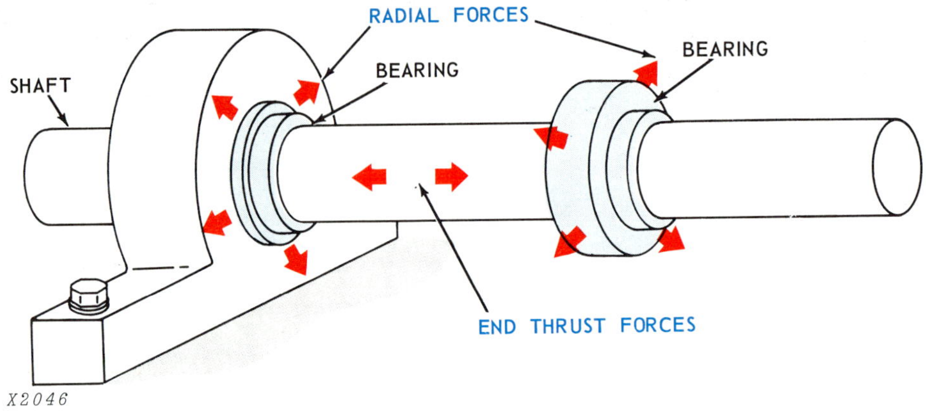

Fig. 29—Load Forces Acting on Bearings

BEARING LOADS

Bearing Loads (Fig. 29) may be of two types:

1) Radial loads—forces perpendicular to the axis of rotation.

2) Thrust loads—forces parallel to the axis of rotation.

Many bearings must be able to carry both radial and thrust loads.

Now let's look at each type of bearing in more detail.

BALL BEARINGS

Ball bearings support a shaft for radial forces as well as thrust forces (Fig. 29). The shaft must be aligned in the bearing bore or the bearing will bind and quickly wear out.

To withstand various radial and thrust forces, the ball bearing types shown in Fig. 30 are widely used. (Thrust forces on each bearing as shown with small blue arrows.) Ball bearings are also available in self-aligning types which compensate for shaft angles in relation to the bearing mount.

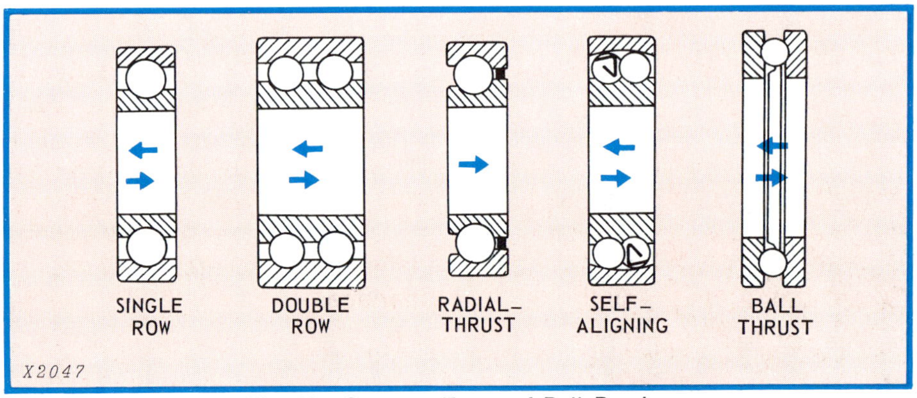

Fig. 30—Common Types of Ball Bearings

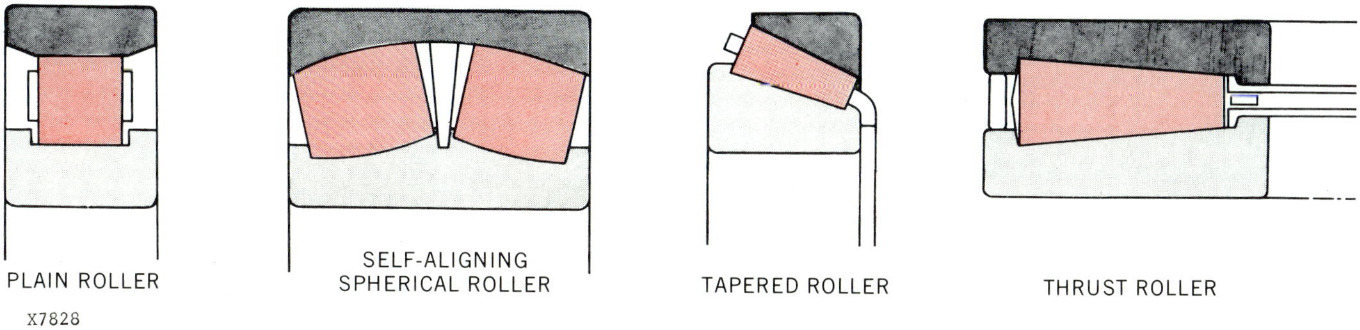

Fig. 31—Common Types of Roller Bearings

ROLLER BEARINGS

Roller bearings are basically the same as ball bearings with the balls replaced by rollers. Often the outer race can be removed without the rollers falling out. Again, the roller bearing can handle radial or thrust forces, or both.

Roller bearings can be designed to handle heavy thrust loads.

For heavy loads—both radial and thrust—*tapered* roller bearings are commonly used. A good example of this is in front wheel bearings. Tapered bearings are also used where high preloads are required on shafts or gears to support thrust.

Fig. 31 shows the common types of roller bearings.

NEEDLE BEARINGS

Needle bearings (Fig. 32) are much like roller bearings except that the rollers are thinner. Also, most needle bearings have no inner race but roll directly on the shaft. The rollers are not separated, but are tightly packed for added support of the shaft.

Needle bearings can support heavy radial loads but must not be placed under thrust loads.

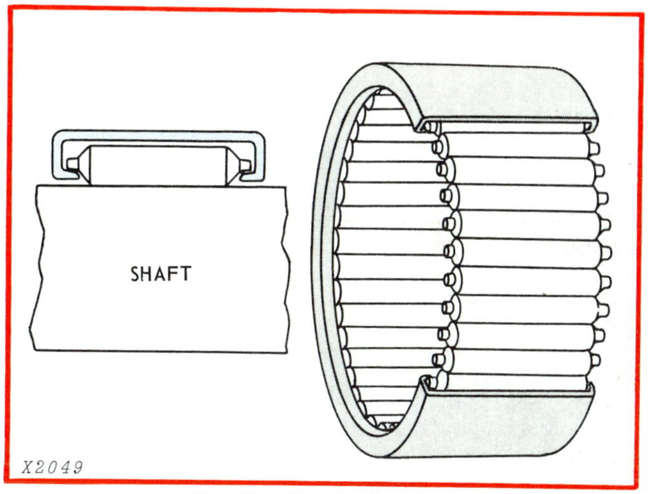

Fig. 32—Needle Bearing

Needle bearings are used in compact locations where relatively high radial loads must be supported. This bearing is frequently located inside a gear which must run free on a shaft or act as an idler. The length of the rollers and their tight packing gives the gear a good support and alignment. Planetary gears are usually supported by needle bearings.

While servicing, if there is doubt about the condition of a needle bearing, replace the bearing.

For more information on bearings refer to FOS-54B Bearings and Seals.

ADJUSTING THE GEAR TRAIN

When a gear train is operated, reaction loads from gears, etc. are transmitted to the bearings and the various parts deflect.

For this reason, the gear train must normally be adjusted for the proper fit between parts.

Three kinds of adjustments are used:

- **Backlash—clearance or "play" between gears in mesh**
- **Endplay—end-to-end movement in a gear shaft due to bearing clearances**
- **Preload—a load within the bearings set up by adjustment**

Let's look at each one and then see how they all work together when adjusting an actual gear train.

BACKLASH IN GEARS

Too much **backlash** in gear trains is the result of: 1) improper mesh between gears or 2) lack of support in bearings.

The result of too much backlash can be broken gear teeth or bouncing of gears under impact forces.

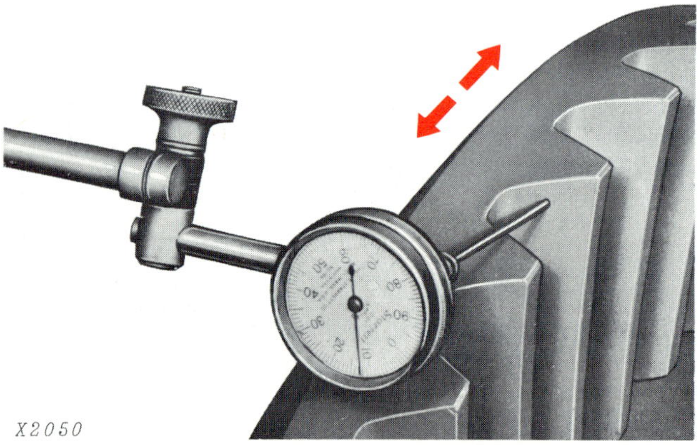

Fig. 33—Checking Backlash On Gear

Backlash is often adjusted to a specified reading on assembled gears (Fig. 33). Move ring gear in toward pinion to reduce backlash, or away from pinion to increase (Fig. 35, step 4).

The dial indicator is mounted so that it registers the full rotary movement of the ring gear shown.

To adjust the backlash reading, shims are often used (see below).

ENDPLAY IN GEARS AND SHAFTS

Endplay refers to a measurable axial looseness of a bearing. Endplay is measured in an unloaded condition.

Preloading is often used to take up the slack and load the bearings (see below).

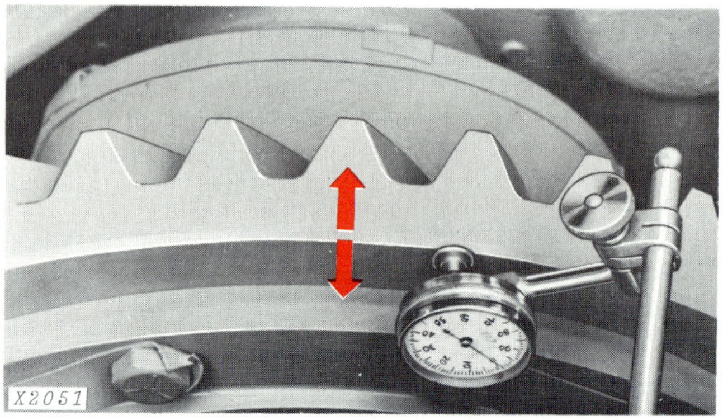

Fig. 34—Checking Endplay On Gears and Shafts

To check endplay (Fig. 34), a dial indicator is mounted against the side of a gear or the end of a shaft. The gear or shaft is then pried in both directions and the readings noted. The difference between the two readings is the endplay.

Shims or adjusting nuts are widely used to adjust endplay.

PRELOADING OF GEAR TRAINS

If the loads are heavy or the thrust too great, gear trains are often **preloaded** to reduce the deflection of parts. This preload must fit the design of the bearings and the strength of the parts.

If bearings are preloaded too tight, they will heat up and fail. If they are set too loose, the supporting parts will deflect too much, causing them to wear rapidly.

Gear trains are preloaded by shims, thrust washers, adjusting nuts, or by using double-race bearings.

Let's use all three of these adjustments—backlash, endplay, and preload—in adjusting an actual gear train.

 Litho in U.S.A.

EXAMPLE: ADJUSTING RING GEAR AND PINION

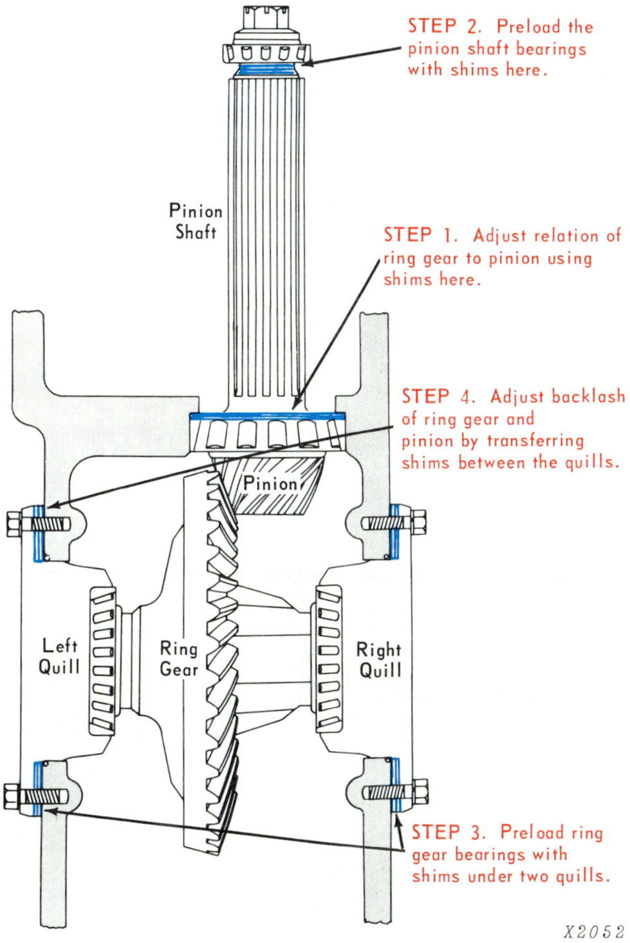

Fig. 35—Adjusting Ring Gear and Pinion (A Typical Example)

A typical ring gear and pinion set is shown in Fig. 35. The pinion shaft at the top meshes with the ring gear at the bottom. The ring gear is supported in two quills which bolt into the sides of the transmission housing. The adjusting shims are shown in blue.

When the gear set is assembled, it is adjusted in four steps:

1. Relation of pinion to ring gear is adjusted with shims under the lower shaft bearing.

2. The pinion shaft bearings are preloaded by using shims under the upper shaft bearing. (The shaft is normally installed with extra shims, endplay is checked, and shims are removed to equal the endplay **plus** the specified preload.)

3. The ring gear bearings are preloaded by shims under the side quills. (Here again the ring gear is normally installed with extra shims, endplay is checked, and shims are removed to equal the endplay **plus** the recommended preload for the bearings.)

4. Backlash between the ring gear and pinion is adjusted by transferring shims under the side quills. Backlash is checked at several points around the ring gear as shown in Fig. 33. To decrease the backlash, shims are transferred from the left to the right quill. To increase backlash, shims are transferred from the right to the left quill. The same shims must be added to the opposite quill as removed from the other one or the preload on the ring gear bearings will be changed.

This is a typical adjustment for a gear train but there are many variations. Consult the machine Technical Manual for the exact steps and readings.

SUMMARY: POWER TRAINS

We have seen how a basic power train works and have looked at the basic elements (Fig. 36).

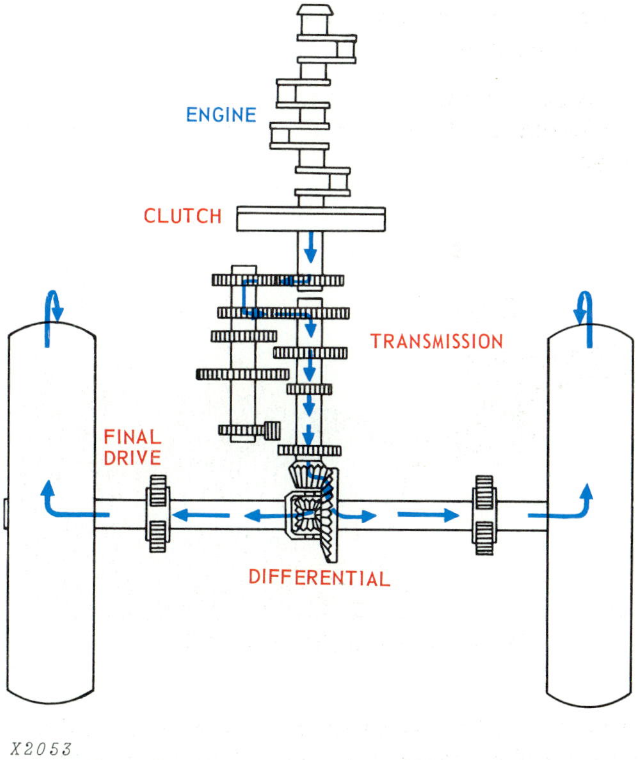

Fig. 36—Parts Of A Machine Power Train

In the rest of this book, we will look at various types of power trains in detail, starting with clutches in Chapter 2.

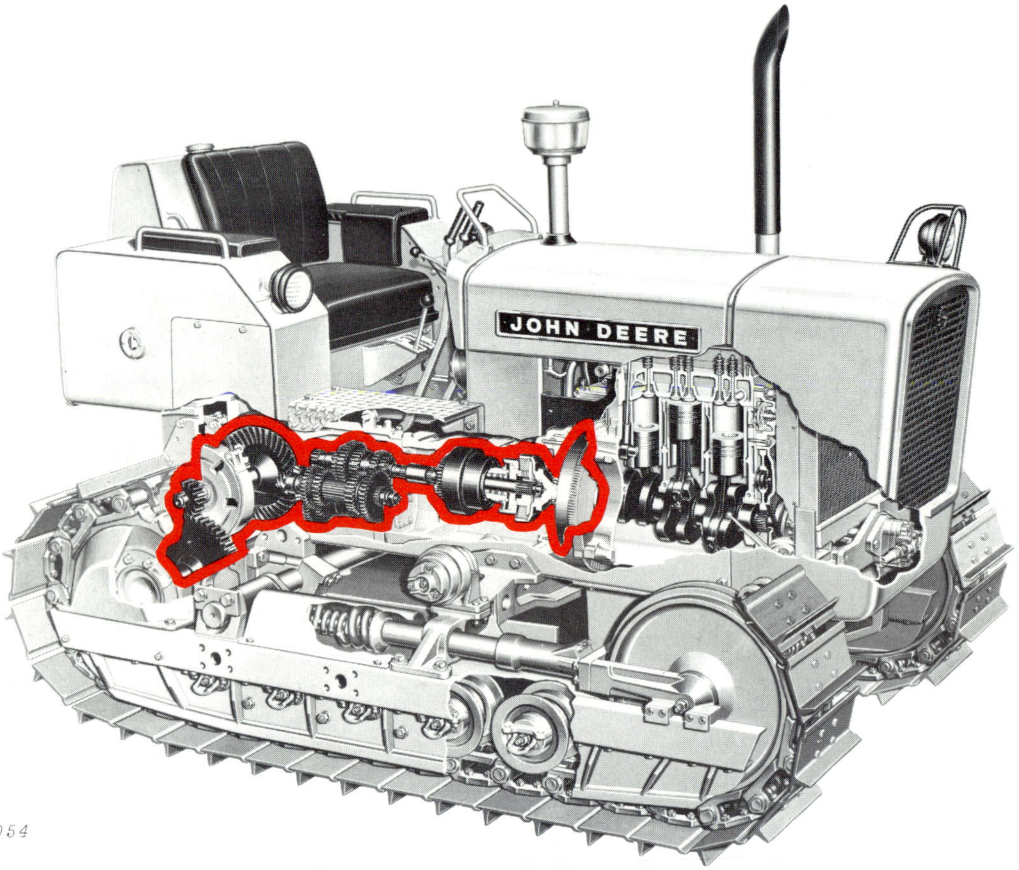

X2054

Fig. 37—Complete Power Train In Industrial Crawler Tractor

X11387

Fig. 38—Complete Power Train In Modern Farm Tractor

 Litho in U.S.A.

Fig. 39—Complete Power Train In Four-Wheel Drive Machine

TEST YOURSELF

QUESTIONS

1. Match each item below with the correct job it performs listed at the right.

a. Clutch	*1. Equalizes power for turning.*
b. Transmission	*2. Connects and disconnects power.*
c. Differential	*3. Selects speeds and direction.*

2. Two gears are in mesh. Gear A has 12 teeth, while gear B has 24 teeth. When gear A has made one complete revolution, how far has gear B traveled?

— 2 revolutions

— ½ revolution

— 4 revolutions

3. True or false? "For speed reduction in low gears, a large gear drives a smaller gear."

4. Does reducing gear speeds increase or decrease the twisting force of the gear shaft?

5. True or false? "When a machine turns, the outside drive wheel must turn faster."

6. What are the three basic ways of transmitting power?

7. Name an example of each type of transmitting power.

8. Name one advantage of helical gear teeth over straight spur gear teeth.

9. Name the two major jobs of bearings in a power train.

10. (Fill in the blank.) "Bearings are often________ to offset the thrust of the power train."

11. (Fill in the blank.) "Looseness in a power train from end to end is called________________."

12. (Fill in the blank.) "Clearance between gear teeth in mesh is called________________."

(Answers on page 9 at end of text.)

CLUTCHES / CHAPTER 2

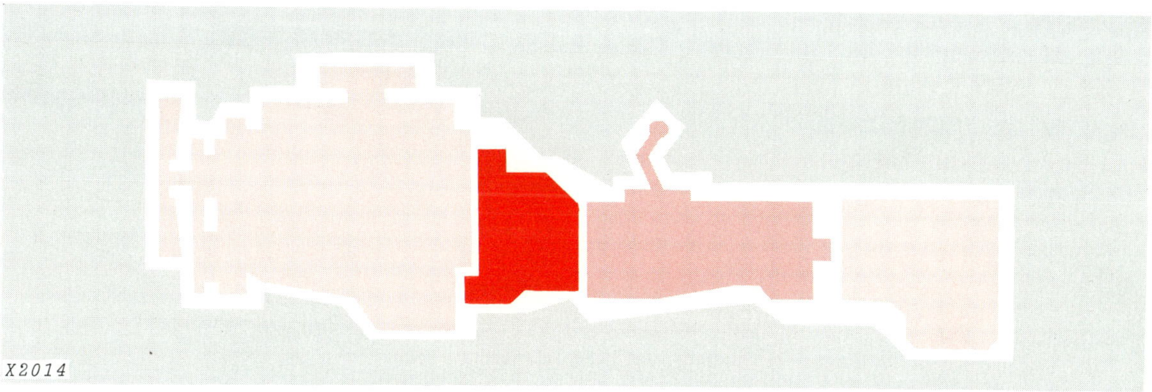

Fig. 1—Clutch In Complete Power Train

INTRODUCTION

The **clutch** is a vital link in the power train. It connects and disconnects power between the source (engine) and the receiver (transmission).

Other clutches connect and disconnect power to auxiliary mechanisms in the gear train itself.

A clutch is used in all machines except some that have hydraulic assist drives or torque converters.

Chapter 1 explained the basic reasons for a clutch. Now we are going to see how the various clutches operate.

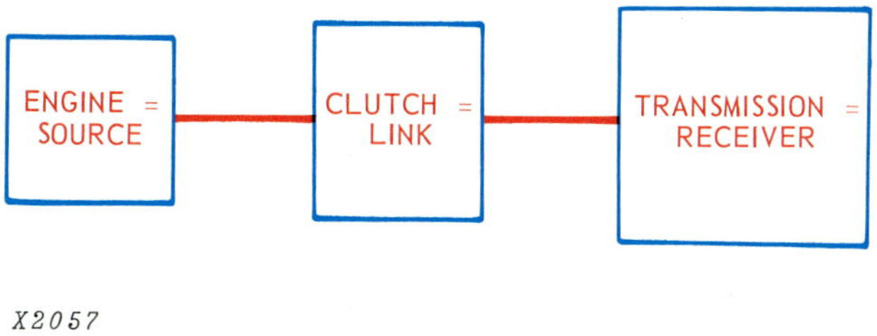

Fig. 2—Relation Of Engine Clutch To Power Train

TYPES OF CLUTCHES

Six types of clutches will be covered in this chapter:

- **Disk and Plate—one or more plates are brought together to transmit torque.**
- **Band—an outside band is tightened over a rotating part to transmit torque.**
- **Overrunning—engagement is allowed in one direction, but the unit freewheels in the other.**
- **Magnetic—a magnetic field holds two parts together, allowing them to rotate as a unit.**
- **Cone—two cone-shaped parts are engaged to transmit torque.**
- **Expanding Shoe—an inner shoe is expanded to contact an outer part to transmit torque.**

Let's look at each.

DISK AND PLATE CLUTCHES

Two major types of *disk* clutches are used:

- **Dry Disk Clutch—operates dry, cooled by air.**
- **Wet Disk Clutch—operates in either oil bath or spray, cooled by oil.**

Let's take a look at each.

DRY DISK CLUTCH (STANDARD)

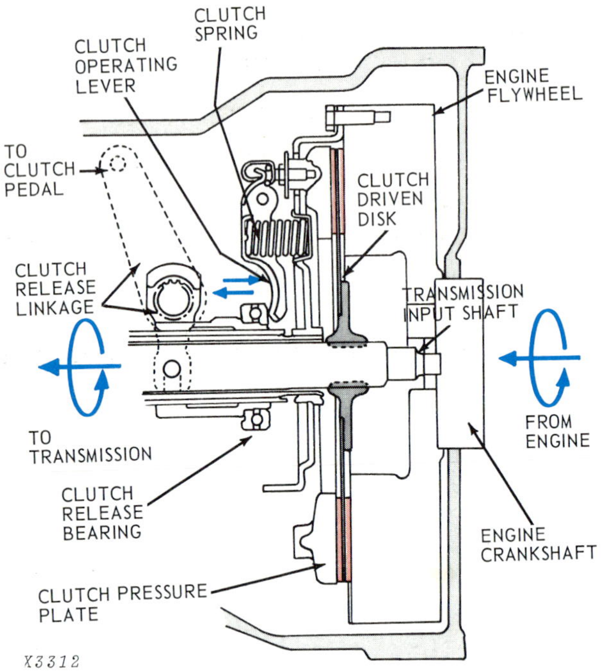

Fig. 3—Dry Disk Clutch (Standard)

The standard dry clutch shown in Fig. 3 has a driven disk with friction facing (red areas) molded to both sides.

The clutch pressure plate housing is bolted directly to the engine flywheel, while the driven disk is splined to the transmission input shaft.

Operation When Disengaged

When the clutch pedal is depressed, the clutch release bearing moves forward against the operating levers (Fig. 3). As the levers pivot, the clutch bolts pull the pressure plate rearward to compress the clutch springs and release the pressure applied to the clutch disk. This separates the driven disk from the flywheel and pressure plate. Power flow is now stopped between the engine flywheel (rotating) and the transmission input shaft (where the driven disk is now rotating freely).

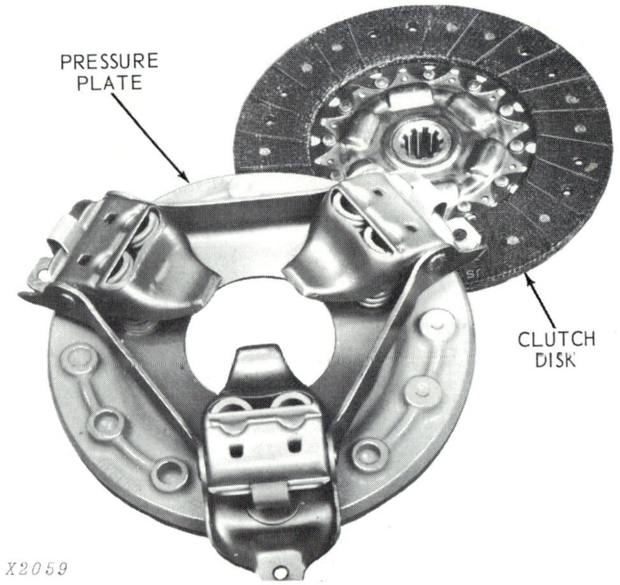

Fig. 4—Dry Clutch Disk and Pressure Plate (Standard)

Operation When Engaged

When the clutch pedal is released, the compressed clutch springs force the pressure plate to move forward to apply pressure on the clutch disk. Since the clutch bolts are secured to the pressure plate, the operating levers force the clutch release bearing rearward. This allows the plate to clamp the driven disk against the flywheel again. Now the flow of power is connected between the engine and the transmission input.

DRY DISK CLUTCH (HEAVY DUTY)

In the heavy-duty disk clutch, both driving and driven members are used.

The **driving** parts are the flywheel, pressure plate, and possibly an intermediate plate.

The **driven** parts are the clutch disks and the clutch shaft of the transmission.

A release bearing is used to disengage the clutch, and spring pressure holds the clutch engaged.

Operation

The clutch shown in Fig. 5 is engaged. Notice that the release bearing is away from the release levers. In this position, the clutch springs exert full pressure on the clutch plate, holding the clutch disk against the flywheel.

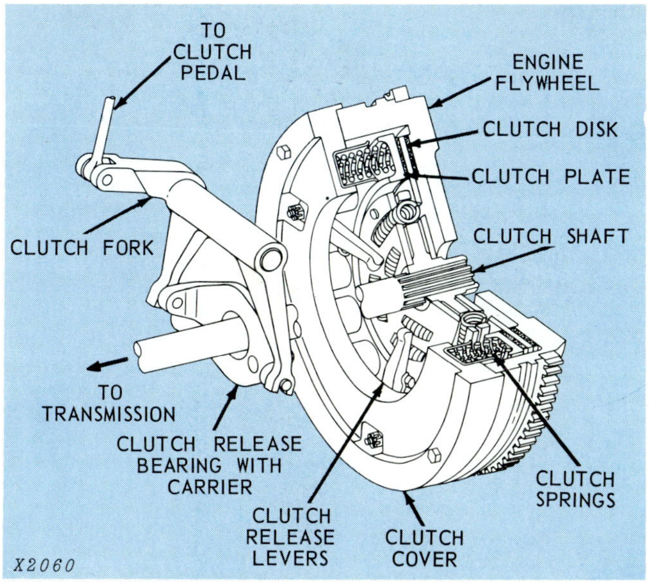

Fig. 5—Heavy Duty Disk Clutch (Engaged)

To disengage the clutch and interrupt the flow of power, the release bearing is pushed forward by the clutch fork. It contacts the release levers, causing them to pivot and pull the clutch plate back away from the clutch disk. The driving parts now rotate without moving the clutch disk and power is interrupted.

Now let's look at the individual parts of a heavy duty disk-type clutch (Fig. 6).

Flywheel

The engine flywheel usually provides a machined surface for the clutch to contact (see Fig. 5).

In some cases a special plate is bolted to the flywheel to provide this surface.

The purpose of the flywheel is:

1) As a balancer for the engine, it levels out power impulses.

2) It provides a surface from which the clutch can pick up torque for transfer to the drive mechanisms.

3) It provides a ring gear for the engine starting motor to contact.

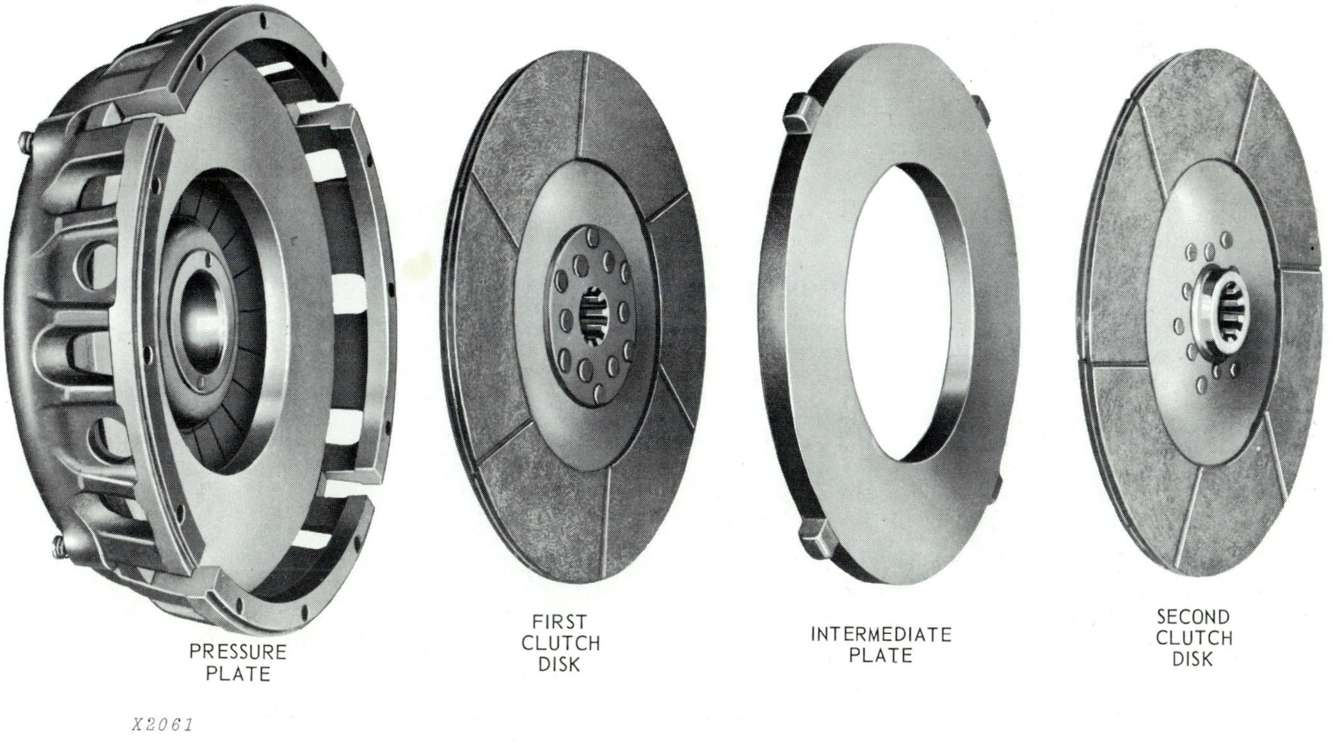

Fig. 6—Parts Of A Disk Clutch (Heavy Duty)

Litho in U.S.A.

Pressure Plate Assembly

The pressure plate engages and releases the clutch disk. A typical plate assembly has a clutch plate, bracket, release levers, and springs.

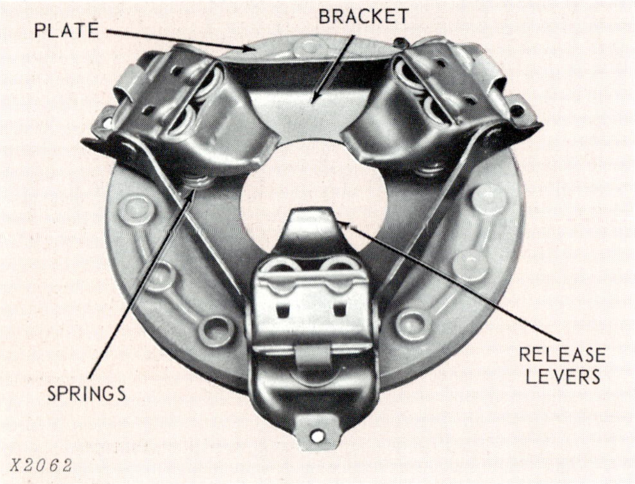

Fig. 7—Pressure Plate Assembly (Heavy-Duty)

The heavy *plate* actually engages the clutch by pressing the disk against the flywheel surface.

The *bracket* mounts the *release levers* and *springs.* The levers pivot in bosses on the plate so that when the levers are pressed down by the release bearing, the plate is lifted from the clutch disk.

The *springs* under each release lever push the plate down again when the clutch is engaged and hold the disk solid. This happens when the force on the levers is released.

Some clutches also have centrifugal weights which add extra pressure for holding the clutch plate against the clutch disk.

Intermediate Plate

The intermediate plate is used in heavy-duty clutches having two disks and is placed between the disks.

When the clutch is engaged, the first clutch disk is held between the clutch pressure plate and intermediate plate, and the second clutch disk is held between the intermediate plate and the flywheel.

When disengaged, the intermediate plate, flywheel and pressure plate assembly rotate as a unit. The clutch disks which are not in contact remain idle.

Clutch Disks

There are two kinds of clutch disks—**rigid** and **flexible** (Fig. 8).

A *rigid* or solid clutch disk is a solid circular disk fastened directly to a center splined hub.

The *flexible* clutch disk is easily recognized by the torsional dampener springs located in circles around the center hub. The purpose of the flexible-type disk will be defined more clearly under the "Dampener" section of this chapter, page 2-15.

Asbestos wire-woven material is the most common lining for these clutch disks. A more recent product is ceramic material, a clay and metal substance which is oven-baked.

A lining must be long-wearing, heat-resistant, and still able to overcome slippage.

Linings are attached to the clutch disk by bonding or riveting.

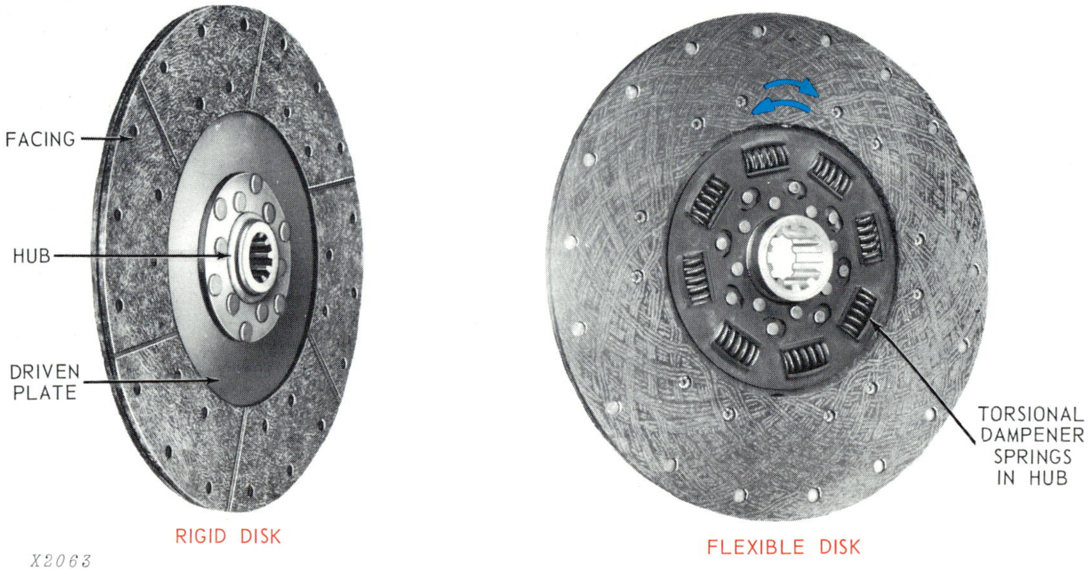

Fig. 8—Heavy Duty Clutch Disks

 Litho in U.S.A.

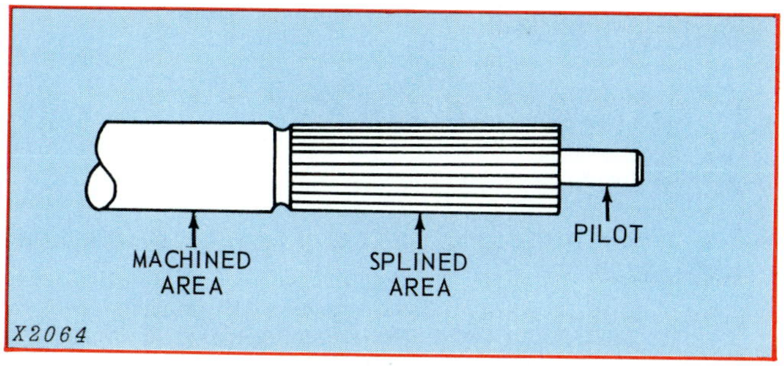

Fig. 9—Clutch Shaft

Clutch Shaft

The clutch shaft projects from the front of the transmission (Fig. 9).

Most clutch shafts have a smaller shaft or pilot which projects from the front end. This pilot rides in the pilot bearing in the engine crankshaft flange.

The splined area is provided for the clutch disk. It must allow the clutch disk to move laterally along the spline but must not allow a rotary rocking motion.

A machined area between the transmission front cover and the spline mounts the release bearing carrier.

A thrust-type ball bearing is the most common release bearing. It rests on this area of the shaft when the clutch is engaged.

SERVICING DRY DISK CLUTCHES

1. Always check the clutch release and pressure mechanisms. Pressure should be even all around the clutch. The release levers must be adjusted so that contact with the release bearing will be uniform and the pressure plate will release evenly.

CAUTION: Use proper equipment and methods as prescribed by your Technical Manual when dismantling the direct-pressure clutch. Spring pressure must be kept under control and released evenly to avoid personal injury.

2. After rebuilding the clutch, check the clutch pedal *free travel* (Fig. 10). When the clutch is engaged, a retracting spring pulls the pedal back so that the release bearing fork does not contact the release bearing. Another spring pulls the release bearing carrier back so that the bearing does not contact the clutch plate levers. Thus, in normal driving, with the clutch engaged and

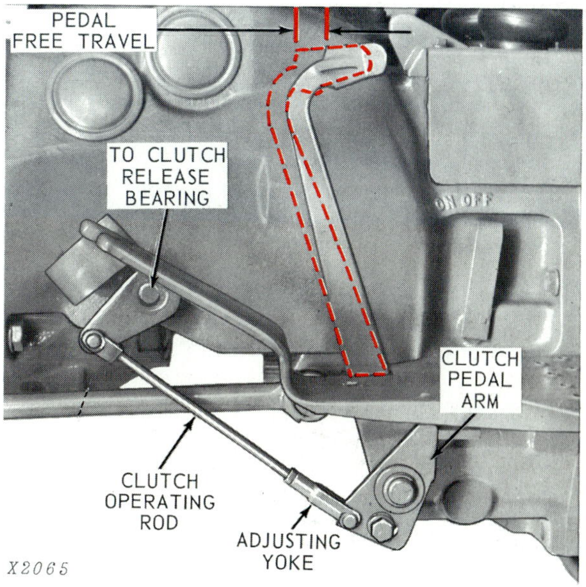

Fig. 10—Clutch Pedal Free Travel Adjustment

the driver's foot removed from the clutch pedal, the clutch release bearing is not turning. This is assured by having the correct free travel on the pedal (Fig. 10).

3. Always check the clutch release setting during clutch service. As the clutch disk facings wear, the clutch disk becomes thinner. The clutch plates then move closer to the flywheel in order to engage the disk. As the pressure plate moves closer to the flywheel, the release fingers move closer to the release bearing. After long operation, these fingers will contact the release bearing and it will run all the time.

Additional wear on the clutch disk cannot be taken up by the pressure plate because the release fingers are contacting the release bearing. The clutch will then start to slip and the release bearing will wear out rapidly.

We will discuss the clutch release adjustment under "Manual Actuating Mechanisms" later in this chapter.

LUBRICATION OF DRY DISK CLUTCHES

Do not use too much lubrication on the clutch pilot bearing, release bearing, or release fork. Excessive lubrication can lead to erratic operation of the clutch.

Some release bearings are sealed units that never need lubrication.

 Litho in U.S.A.

USES OF DRY DISK CLUTCHES

Dry clutches are widely used on large trucks and heavy industrial units. The main advantage is in the larger contact area. They are not recommended where frequent disengaging or slipping is required because of heat build-up.

WET DISK CLUTCH

The wet disk clutch operates in an oil bath or spray. Operation is similar to the dry clutch.

Operation

This clutch is either mechanically or hydraulically actuated.

To engage the clutch, a hydraulic or mechanical force is exerted on the rear plate of the clutch pack. As the force increases, more power is transmitted.

Operation in oil provides these advantages: *Reduced wear and cooler operation.*

Disks for Wet Clutches

One major difference between the dry and wet clutch is in the disk facing material. In a wet clutch, this material must grip when soaked in oil.

Crosshatch marks are on the facing surface of the wet disk. Cork was a common lining material in the past. Bimetallic plates without lining are often used now. Bimetallic plates have steel driving disks and bronze driven plates.

Servicing

Consider these points when servicing a wet clutch:

- *Is the oil being circulated properly?*
- *On units equipped with pumps, are the lines clogged?*
- *Is the pump moving its full capacity of oil?*
- *Is the inlet line or filter dirty?*

Before disassembly, check the oil level, and look for leaking oil.

During repair, check all the oil seals at the clutch release forks, flywheel, and clutch shaft end of the housing. Replace worn seals. If a seal has small cracks on the lips or retainer, it should be replaced. If the seal material appears compressed and is glazed, it is wise to replace it.

The best practice with seals is, "If in doubt, throw it out."

Failure to keep oil in the clutch housing will cost a lot more in repairs than the price of an oil seal.

For details, see the machine Technical Manual.

Lubrication

Use the type and grade of oil specified by the manufacturer.

Add oil if necessary to retain the proper level. This will avoid clutch failure from scoring or heat warpage.

Because these clutch disks are both lubricated and cooled by oil, the proper oil supply is vital to normal service life.

BAND CLUTCHES

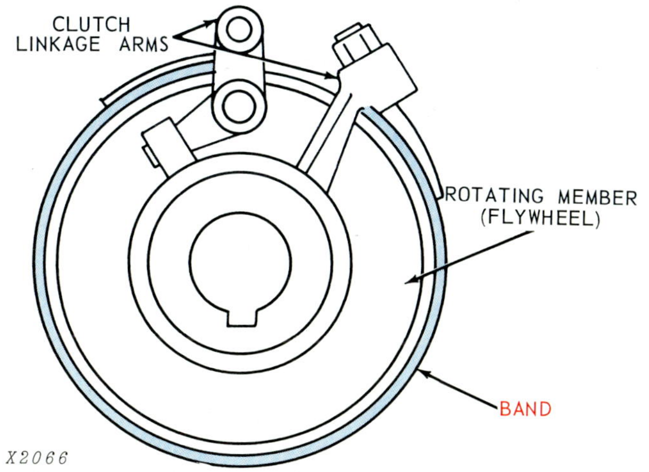

Fig. 11—Band Clutch

The band clutch is different from the others in that it receives its torque from the *outside* surface of a rotating member, not on the side or on the same axis.

OPERATION

To engage, the linkage arms (Fig. 11) are drawn closer together, causing the band to tighten. The band and flywheel then revolve as a unit, transmitting torque.

Advantages include the ability to absorb shock loads, uniform wear of linings, and less heat. A more constant torque is provided because of the large-diameter contact. The lining materials used include leather, wood, and asbestos-woven material.

USES

Band clutches are adaptable to large stationary equipment rather than mobile equipment because of their bulky size.

OVERRUNNING CLUTCHES

An overrunning clutch (Fig. 12) will automatically engage in one direction, but will freewheel in the other.

In the lower picture (Fig. 12), the entire shaft is revolving counterclockwise. As long as the drive line revolves, the clutch will remain **engaged.**

If we stop the power or reverse the rotation, the clutch will overrun or **freewheel** as shown in the lower picture. Engagement will not take place until the drive line attains more speed than the driven line.

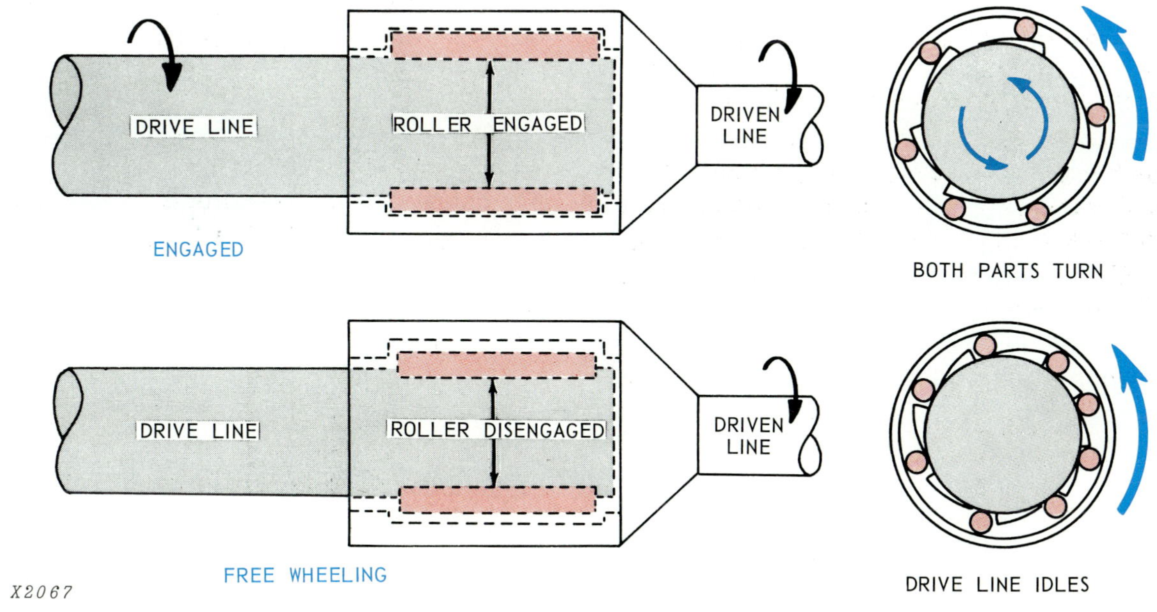

Fig. 12—Overrunning Clutch

TYPES OF OVERRUNNING CLUTCHES

Overrunning clutches have three basic types:

- **Roller**
- **Cam or Sprag**
- **Spring**

Let's look at each one.

Roller Overrunning Clutch

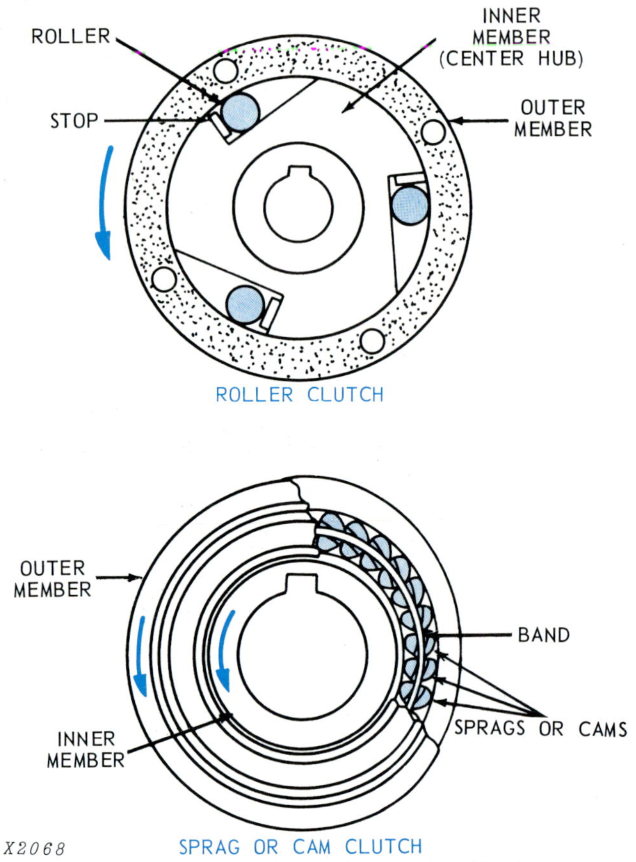

Fig. 13—Roller and Sprag Overrunning Clutches

The upper part of Fig. 13 shows a side view of a roller clutch. Notice that there is an outer member and an inner member. As the center hub turns, the rollers move outward on the inclined ramp until they wedge between the outer and inner members. The members then revolve as a unit to transmit torque.

If we reversed the inner member, the rollers would be free to return down the ramp against the stops. In this position, there would be no contact between the inner and outer member, and they would disengage and freewheel.

Cam Or Sprag Overrunning Clutch

The sprag-type clutch works in much the same manner as the roller.

The lower part of Fig. 13 shows a sprag or cam clutch. Sprags or cams take the place of the rollers.

The clutch is shown in an engaged position.

To *disengage* the clutch or freewheel, the inner member would have to either stop or turn clockwise. This would allow the cams or sprags to lean forward. The band of material passing through the sprags prevents them from lying flat.

To *engage* the clutch, the inner member rotates counterclockwise as shown. This causes the sprags to stand upright, locking them between the inner and outer members. The members held together by the sprags then rotate as a unit, transmitting the torque.

Fig. 14—Sprag Type Clutch

Spring-Type Overrunning Clutch

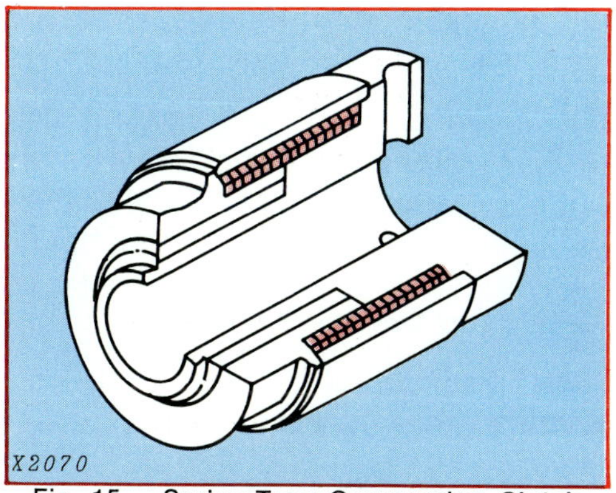

Fig. 15— Spring-Type Overrunning Clutch

The spring clutch is restricted to smaller applications. A coil spring is wrapped lightly around a shaft or collar (Fig. 15). When the spring is rotated in one direction, it tightens on the shaft and drives it. When rotated in the other direction, it loosens and does not drive.

OVERRUNNING CLUTCH SERVICE

Refer to your Technical Manual for information on servicing and lubricating these units. Failure to freewheel defeats the purpose of this clutch and can result in costly damage.

MAGNETIC CLUTCHES

There are two types of magnetic clutches:

- **Direct**
- **Indirect**

In the DIRECT, a magnetic field is set up which draws the driving and driven plates together.

In the INDIRECT, a magnetic field is set up between the inner and outer members, which draws fine metallic particles to form a solid connection between the members. The greater the mass of particles, the greater the transfer of torque.

OPERATION

The operation of magnetic clutches will be covered in detail under "Electrical Operating Mechanisms" later in this chapter.

USES

Magnetic clutches are used where frequent engaging and disengaging is required. This electrical clutch gives either positive drive or slippage which cushions any shock loading.

CONE CLUTCHES

The cone clutch is made up of two cone-shaped members (Fig. 16).

One member provides an internal machined surface for transferring torque and serves as a drive member. The other member provides an external lined surface for picking up the torque and serves as a driven member.

The material used for lining is usually woven asbestos bonded to the cone surface.

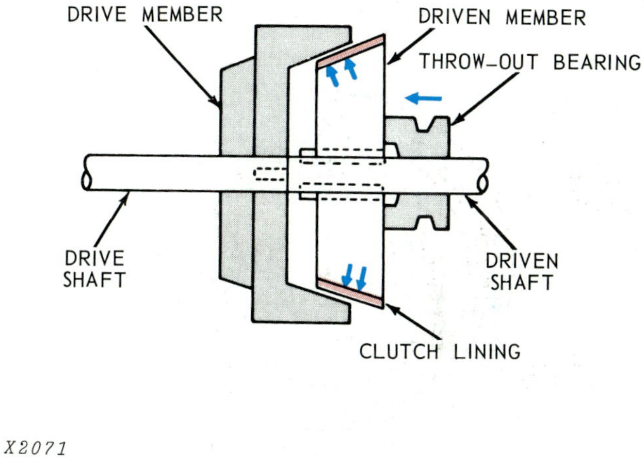

Fig. 16—Cone Clutch

OPERATION

The cone clutch is engaged as follows:

As the throw-out bearing moves on the driven shaft, it slides the driven member forward (Fig. 16). The outer lining of the driven member contacts the inner surface of the revolving drive member to transmit the torque.

USES

The cone clutch was at one time used in trucks and autos, but now is used more for machine auxiliary power trains.

Litho in U.S.A.

EXPANDING SHOE CLUTCHES

The expanding shoe clutch can be operated either *mechanically* or *centrifugally* (Figs. 17 and 18).

OPERATION

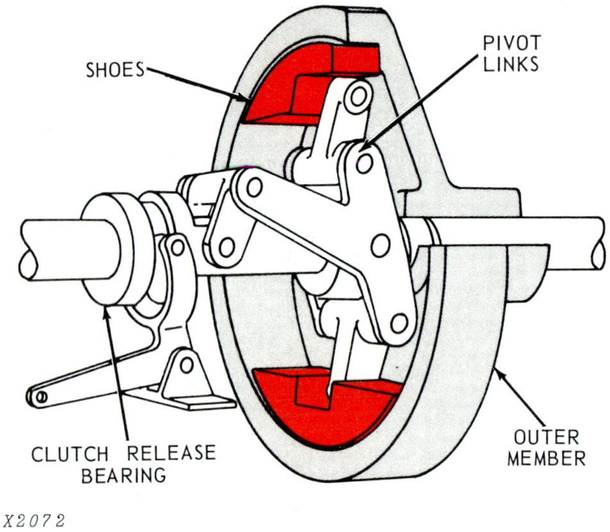

Fig. 17—Expanding Shoe Clutch (Mechanical Type)

To engage the **mechanical** shoe clutch, a release bearing is pushed forward (Fig. 17). This pivots the links and expands the shoes out against the outer member.

Moving the bearing back releases the pressure on the shoes, disengaging the clutch.

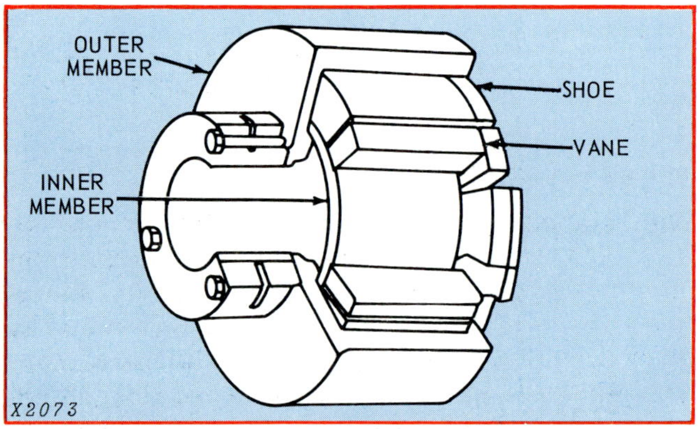

Fig. 18—Expanding Shoe Clutch (Centrifugal Type)

The **centrifugal** clutch (Fig. 18) has an inner member with vanes spaced around the hub. The outer member closes the end of the channels formed by the vanes. Shoes are placed in the channels and are free-floating.

When the inner member rotates at a high speed, the shoes move outward and contact the outer member. The faster the inner member rotates, the greater the centrifugal force holding the shoes against the outer members. The vanes help to hold the shoes and thus form a somewhat solid connection between the two members.

The clutch disengages automatically once the rotating drive member slows down.

OTHER TYPES OF CLUTCHES

We have discussed mechanically- and electrically-operated clutches. New clutches are available using **pneumatics** (air) or **hydraulics** (oil) for control.

PNEUMATIC CLUTCHES

The air-actuated clutch is a clutch pack which contains a small rubber tube. This tube is positioned behind a driven member and a fixed member.

To engage the clutch, a valve is opened and air is forced into the tube, causing it to expand. This expansion forces the driven plate against the drive plate. The higher the air pressure, the greater the force exerted by the tube against the driven plate. Air pressure is decreased to disengage the clutch.

This clutch provides smooth engagement because the rubber tube absorbs the shocks.

HYDRAULIC CLUTCHES

Hydraulic clutches work in much the same way as the pneumatic clutch, but use oil instead of air.

Oil is forced into a chamber between the fixed and moving driven members. A valve is then opened, allowing oil to flow into the chamber. As oil pressure builds up, the walls of the chamber are forced outward, causing contact between the driven and drive members. The amount of pressure exerted on the clutch disk is regulated by the pressure of the oil in the chamber.

These clutches are also fairly shock resistant.

 Litho in U.S.A.

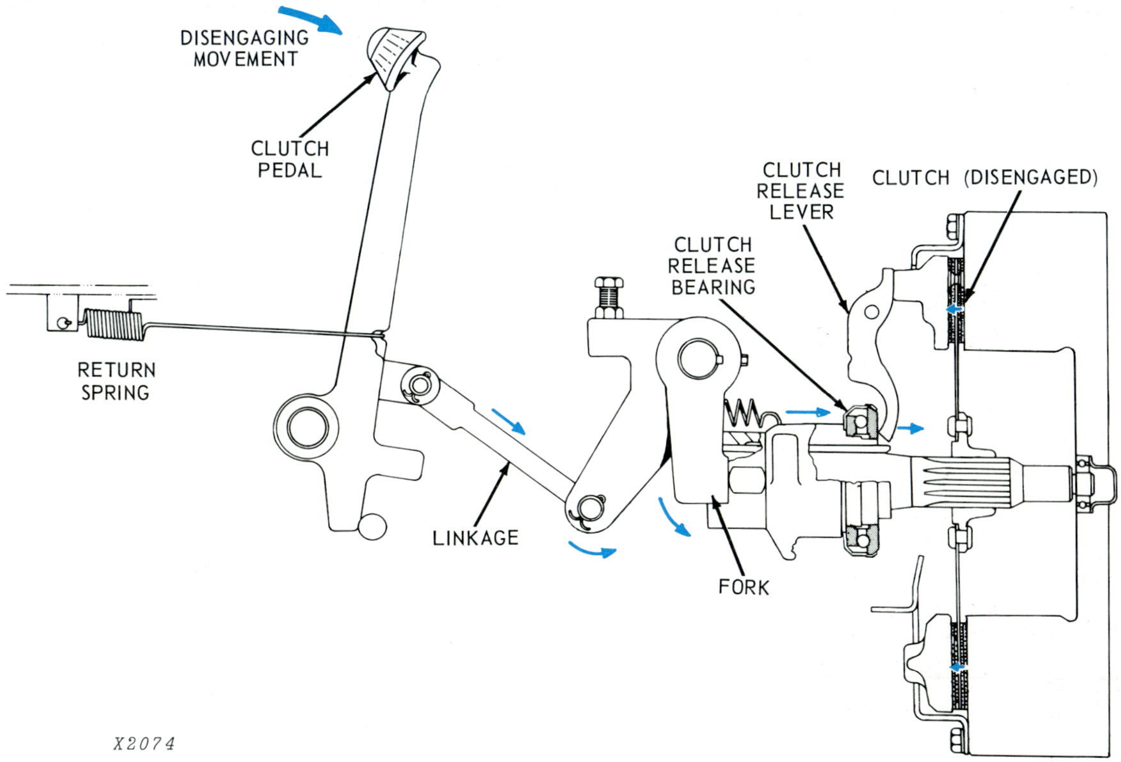

Fig. 19—Standard Mechanical Clutch Linkage

SLIP CLUTCHES

Another type of clutch is the **slip clutch.** Its purpose is to allow slippage at moments of excessive demand or shock. It serves as a safety device in power trains.

This clutch is covered in detail in Chapter 10 of this manual.

CLUTCH OPERATING MECHANISMS

Clutches can be operated in three major ways:

- **Mechanically**
- **Hydraulically**
- **Electrically**

We have also mentioned pneumatics, or air, but let's concentrate on the three controls listed above.

MECHANICAL CLUTCH CONTROLS

Mechanical operation of clutches is the using of linkages and springs to help in manually engaging and disengaging. There are two basic types:

1) Standard

2) Over-Center

STANDARD LINKAGE is the most common; the clutch remains engaged unless a pedal is pushed and held.

OVER-CENTER LINKAGE is where mechanical advantage is used to lock the clutch engaged or disengaged.

Standard Linkage

With standard linkage (Fig. 19), foot motion on the clutch pedal moves through the linkage to the clutch release bearing fork. This pushes the release bearing against the clutch release levers, releasing or disengaging the clutch. Some clutches use springs which serve as aids in disengaging and also as return springs when the clutch is engaged.

Over-Center Linkage

Fig. 20 shows an expanding shoe clutch with an over-center linkage to lock it in place while engaged or disengaged.

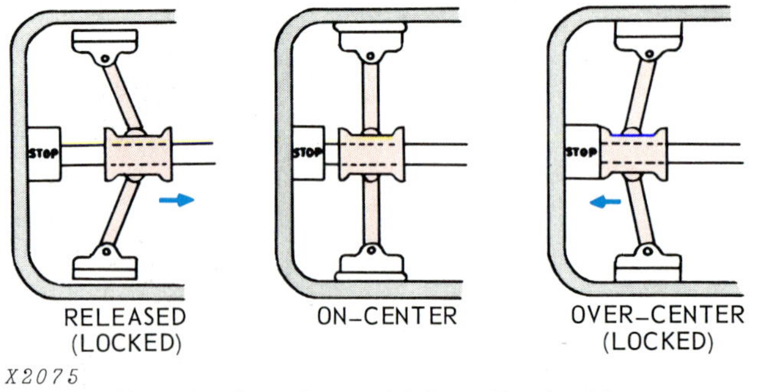

Fig. 20—Over-Center Linkage For Locking The Clutch While Engaged or Disengaged

When the center collar is moved forward, the two arms attached to it move outward. An *on-center* position is obtained when the arms are in line with each other. This is the point of maximum pressure.

If we move the collar ahead still further against the stop, the arms are in an *over-center* position and some of the pressure is relaxed. The collar cannot move ahead because it is against the stop. Rearward movement is impossible unless assistance is provided to overcome the high pressure at the on-center position.

This type of mechanism has the advantage of being able to lock either engaged or disengaged without keeping a grip on the lever or pedal.

Service of Mechanical Clutch Controls

When repairing the clutch, always examine the linkage.

Worn bushings, bent rods, broken springs and damaged cotter pins can result in excessive force being required to operate the clutch.

Clutch pedal free travel is often used as a guide to conditions inside the clutch.

The actual free pedal is more easily determined if checked by hand and should be the clearance between the clutch release yoke fingers and the release bearing housing.

Clutch pedal free play varies; set it according to Technical Manual instructions.

Whenever free pedal travel falls below ½ inch (13 mm), the clutch should be adjusted internally. If free pedal travel is not then restored to about ½- to 1¾-inches (13-44 mm), readjustment of the external linkage may be necessary.

Check the release yoke movement to verify that free pedal travel is actually the release bearing clearance, and not lost motion in linkage due to worn clevis eyes and pins, or worn bushings and shafts at the clutch pedal or linkage arms.

HYDRAULIC CLUTCH CONTROLS

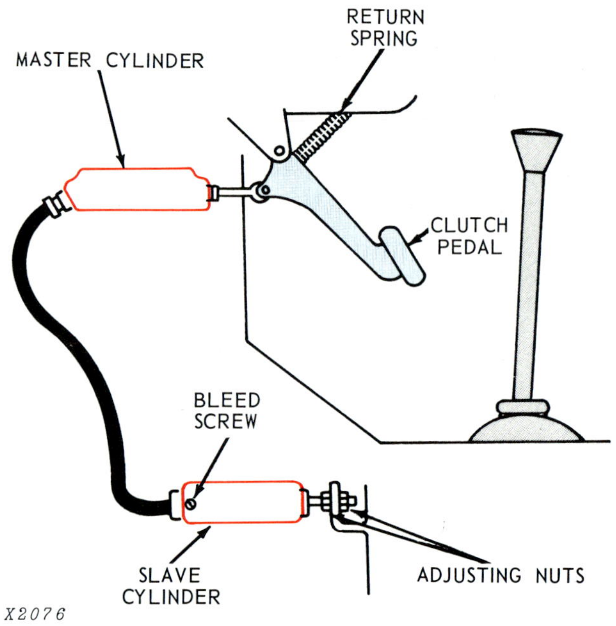

Fig. 21—Hydraulic Clutch Controls

 Litho in U.S.A.

A **master** cylinder similar to the brake master cylinder is attached to the clutch pedal (Fig. 21). A **slave** cylinder is connected to the master cylinder by flexible pressure hose or metal tubing. The slave cylinder is also connected to the clutch release yoke lever.

Movement of the clutch pedal actuates the clutch master cylinder. This movement is transferred by hydraulic pressure to the slave cylinder, which in turn actuates the clutch release yoke lever, disengaging the clutch.

Uses

Hydraulic clutch controls are used on large units which require high pressure to disengage.

By using hydraulics, a small force can be multiplied and used effectively to disengage the clutch.

Service

1. Adjust the clutch pedal free travel per the Technical Manual.

2. Watch for leaking seals, low fluid level, broken or kinked lines, and leaking pistons in the cylinders.

3. Always bleed the air from the system after repair or replacement of parts. In most designs, a special screw is provided to allow bleeding the air from the system.

4. Check the fluid level in the reservoir and add oil if necessary after bleeding the system.

5. Adjust the slave cylinder rod to set the clutch throwout bearing at the proper position.

ELECTRICAL CLUTCH CONTROLS

Electrical clutch controls may be one of two types:

- **Direct-Action**
- **Indirect-Action**

Let's look at each one.

Direct-Action Controls

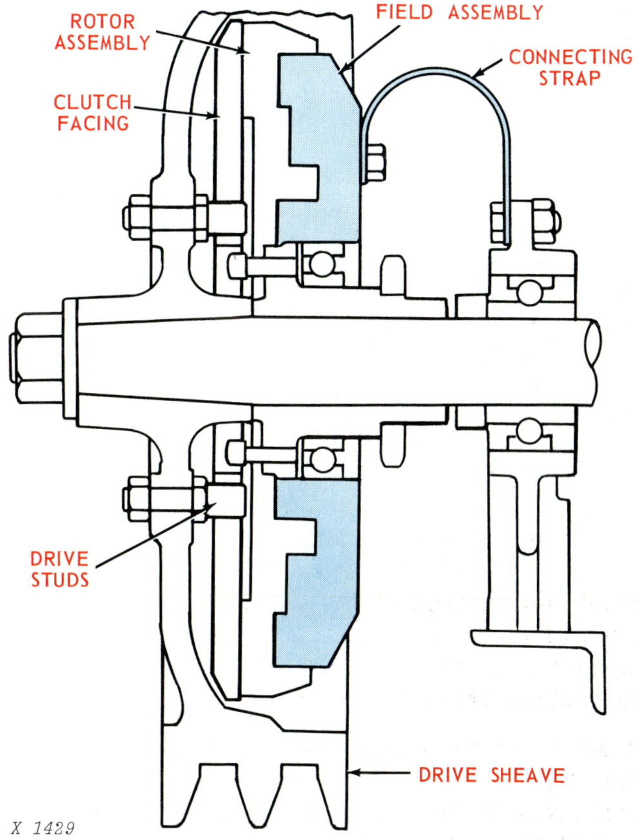

Fig. 22—Direct-Action Electrical Control For Clutch

The direct-action clutch consists of a field coil assembly, rotor unit, face plate, condenser and operating switch (Fig. 22).

When the switch is actuated, current through the field coil inside the rotary assembly sets up a magnetic field which draws the clutch facing against the facing on the rotor assembly. (The face plate is free to slide on the drive studs of the mechanism). Power is then transmitted from the drive pulley through the face plate, rotor, and hub to the drive mechanism.

When the switch is operated again, the magnetic field is collapsed. The face plate is freed from the rotor and the transmitting of power to the mechanism is instantly stopped.

An air-conditioner compressor clutch is an example of a direct action clutch. For more information on air conditioning compressors and clutches, see FOS "Air Conditioning" and Technical Manual.

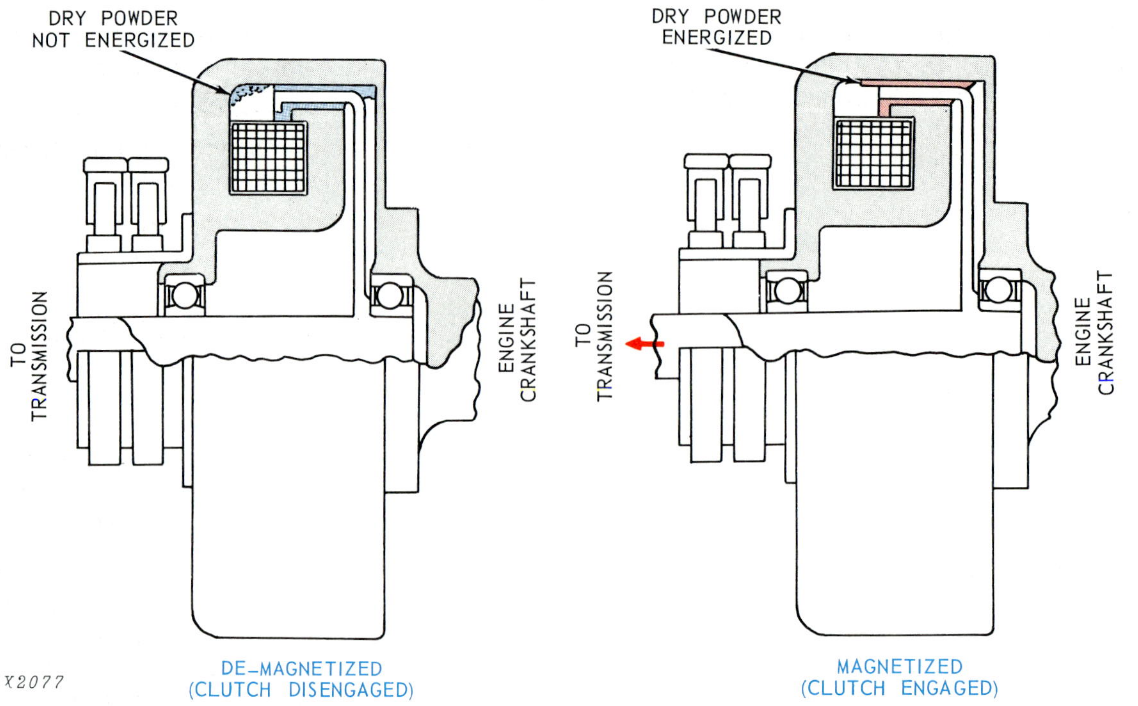

Fig. 23—Indirect-Action Electrical Control For Clutch

Indirect-Action Controls

This clutch consists of a driving member containing a coil, an inner driven member, and a mixture of magnetizable metal and dry lubricant (Fig. 23).

When the coil is energized, the fine metal particles mixed with dry lubricant are driven together in a strong bond which transmits the load. The mixture is literally "frozen" and "unfrozen" in direct proportion to the amount of current applied.

Any desired amount of slippage or solid drive can be obtained by increasing or decreasing the current flow.

Service Of Electrical Clutch Controls

If the clutch suddenly loses power and fails to function, look for a failure in the electrical circuit.

First check the electrical connections, wires, switch and the circuit breaker in the switch.

If these components are okay, check the voltage to the field and amperage of the field.

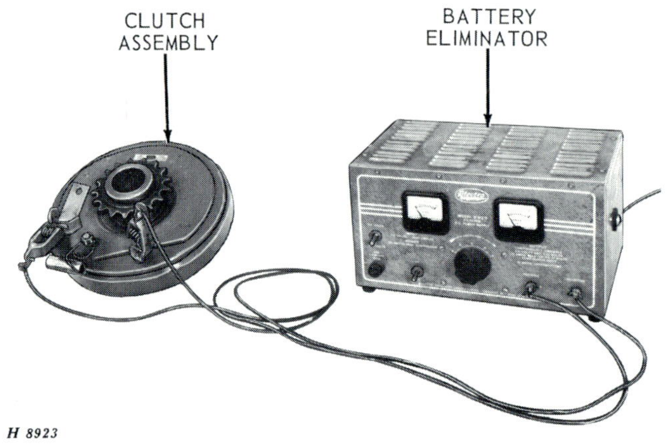

Fig. 24—Testing Field Coil In Electrical Clutch

To check the voltage to the field, the amperage of the field or the resistance of the field, use a battery eliminator (Fig. 24) or a storage battery and voltohmeter. For testing details, see the FOS "Electrical Systems" manual, Chapter 8.

Check the specifications on each clutch for correct input, amperage, and resistance per the machine Technical Manual.

DAMPENERS

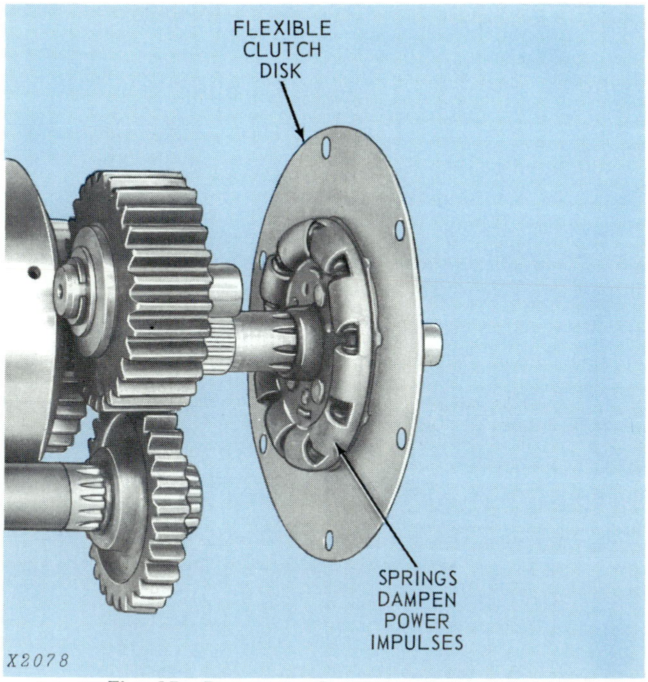

Fig. 25—Dampener Built Into Clutch Disk

The dampener is a shock-absorbing feature built into a flexible clutch disk (Fig. 25). The primary purpose of a flexible disk is to absorb power impulses from the engine which would otherwise be transmitted directly to the gears in the transmission.

A flexible clutch disk has torsion springs and friction disks between the plate and hub.

When the clutch is engaged, the springs cushion the sudden loading by flexing and allowing some twist between the hub and plate.

When the "surge" is past, the springs release and the disk transmits power normally.

MAINTENANCE OF CLUTCHES

Here are some things to watch for when servicing clutches:

1. Use care when disconnecting the clutch linkage. A bent rod or damaged carrier can result in lost motion, hard operation, or faster wear.

2. Don't let the full weight of the transmission rest on the clutch shaft. Use the proper hoists or jacks and keep the parts aligned when separating a machine. Failure to do so will result in damage to the clutch disk and release bearing.

3. Mark the pressure plate assembly and flywheel before removing the pressure plate. When assembled, the unit can be matched with the marks and engine balance maintained. When replacing the entire unit, clutch balance is not so critical because the new unit will be static-balanced. Rear crankshaft oil seal failure is often the result of an unbalanced clutch.

4. After removing the pressure plate assembly, check these four areas: clutch plate, clutch disk, flywheel plate and release bearing.

CAUTION: Use proper equipment and methods prescribed by the Technical Manual when dismantling the pressure plate assembly. Spring pressure must be kept under control and released evenly to avoid personal injury.

CLUTCH PLATE MAINTENANCE

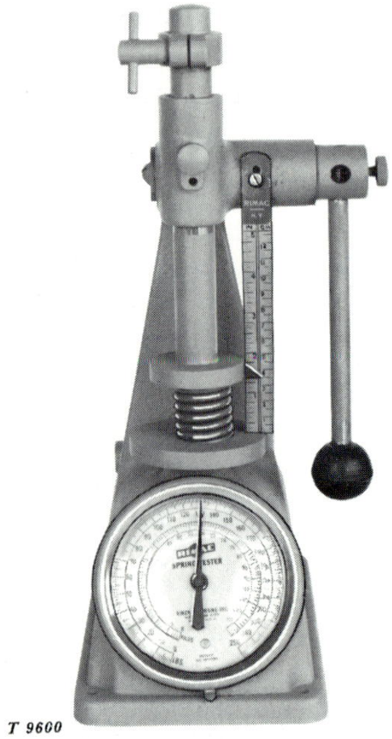

Fig. 26—Checking Tension of Clutch Springs

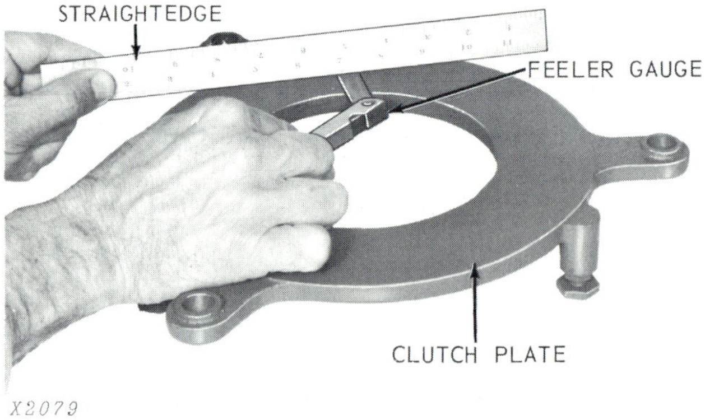

Fig. 27—Checking Clutch Plate For Warping

The clutch plate must be free from ridges, score marks and rough spots. Check the spring tension in the pressure plate assembly (Fig. 26) to make sure the plate is contacting evenly. Watch for excessive wear on release levers. Replace weak or broken springs. If the clutch plate is warped (Fig. 27), replace it.

CLUTCH DISK MAINTENANCE

The clutch disk is actually the heart of the clutch. Check for high and low spots of wear on the friction facings (Fig. 28). On riveted linings, use the level of material above the rivet head as an indicator of wear. Replace the linings if the rivet heads are flush or just below the outer surface.

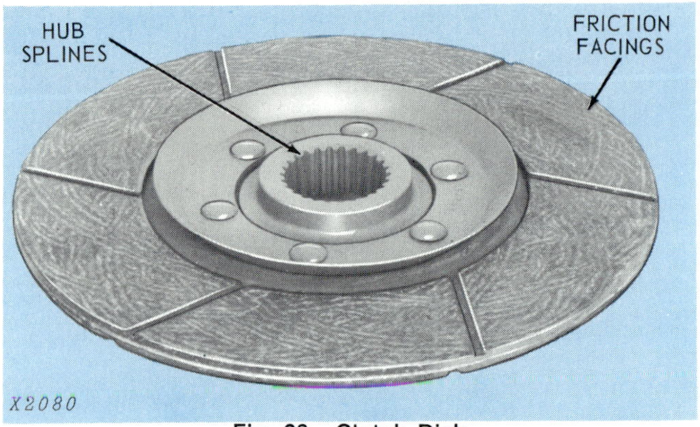

Fig. 28—Clutch Disk

Replace the disk if it appears glazed or cracked. Check any cushion springs and replace if damaged.

Make one last check of the clutch disk: Slide the disk onto the clutch shaft of the transmission. It should slide easily without restriction. If the disk will rock back and forth or from side to side, replace it. This may mean that the splines are badly worn. Try a new disk on the clutch shaft. If you get the same motion, the clutch shaft is worn and it will have to be replaced.

A good indication of clutch alignment is the clutch disk hub. If there is uniform wear along the splines, this means the clutch is aligned. If there is excessive wear on the front of the spline, this means the clutch is misaligned.

FLYWHEEL MAINTENANCE

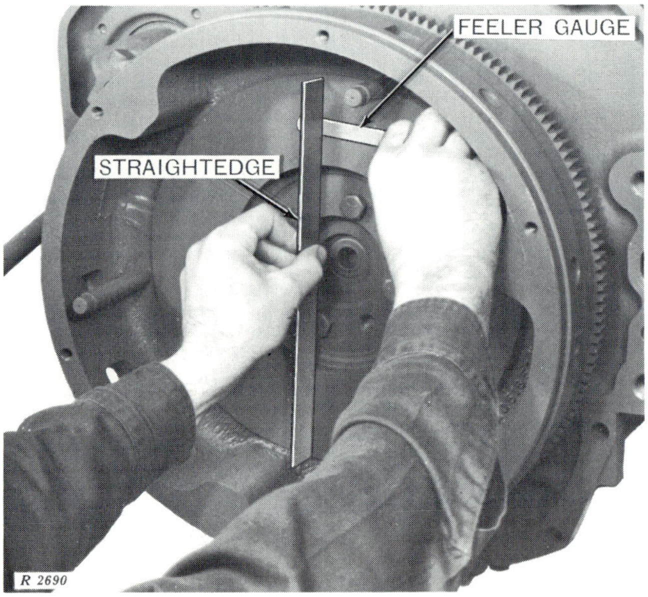

Fig. 29—Checking The Flywheel Friction Surface For Roughness

The engine flywheel surface should be relatively free from roughness and ridges to engage the clutch correctly. Check it with a straightedge and feeler gauge (Fig. 29).

Look for heat checks on the flywheel surface. If the check markings are excessive, the flywheel must be replaced.

If the pilot bearing for the transmission input shaft appear out-of-round or badly worn, it should be replaced. Score marks on the pilot shaft indicate a bearing failure.

CLUTCH RELEASE BEARING MAINTENANCE

Check the release bearing to see if it rotates freely. If in doubt about the wear or lack of lubrication, replace this bearing.

SUMMARY: CLUTCH MAINTENANCE

When the clutch is disassembled for repair, take a few minutes to inspect all associated parts. This could prevent future problems and expense.

A rule of thumb for clutch repairs might be, "If in doubt, replace." Consider what is involved in the job. Will it be worth saving a few dollars now and having to do the entire job again in a few months?

We have given you the basic things to look for, but they are not the complete story. Remember these points and use your machine Technical Manual as a further guide for each clutch.

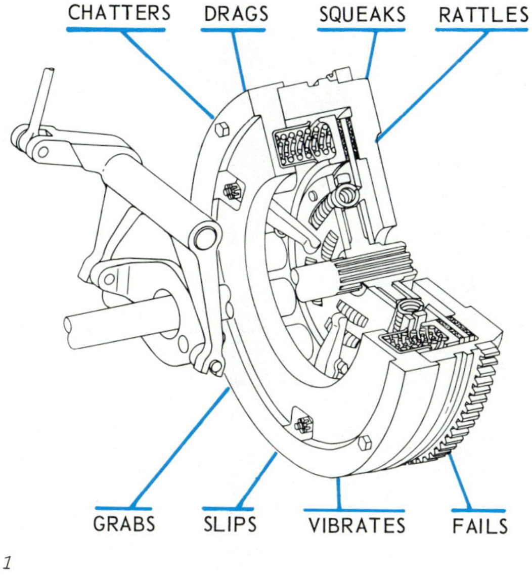

Fig. 30—Clutch Troubles

TROUBLESHOOTING OF CLUTCHES

The eight basic clutch troubles are:

- **Chattering:** *especially in low or reverse speeds.*
- **Dragging:** *or failure to release promptly and fully, thus making gears hard to shift.*
- **Squeaks:** *particularly when pedal is depressed.*
- **Rattles:** *especially at low speeds or standing.*
- **Grabbing:** *violent and sudden engagement.*
- **Slipping:** *failure to transmit full power.*
- **Vibrations** *at either high or low speed or periodically.*
- **Failure** *to transmit power at all.*

Note that different clutch problems can have the same cause. For example, oil or grease on the clutch disk facings can cause the clutch to *slip,* but may also make it *chatter, drag,* or *grab.*

CHATTERING

1. Oil or grease on clutch disk facings.
2. Glazed or worn facings.
3. Worn, loose or spongy engine mountings.
4. Worn or loose splines in clutch hub or on clutch shaft.
5. Wear or looseness in universal joint, differential, or drive axles.
6. Splined disk hub sticking on splined shaft.
7. Cracked or scored pressure plate or flywheel face.
8. Warped clutch disk.
9. Warped pressure plate.
10. Pressure plate sticking on driving studs.
11. Sticking or binding release levers.
12. Unequally adjusted release levers.
13. Unequal length or strength of clutch springs.
14. Bent clutch shaft.
15. Misalignment of power train.

DRAGGING

1. Oil or grease in clutch.
2. Warped clutch disk.
3. Splined disk hub sticking on splined shaft.
4. Sticking pilot bearing or bushing.
5. Sticking release sleeve.
6. Warped pressure plate or clutch cover.
7. Broken disk facing.
8. Accumulation of dust in clutch.
9. Incorrect clutch or pedal adjustment.
10. Engine idling too fast.
11. Misalignment of parts.

SQUEAKS

1. Clutch release bearing needs lubrication.
2. Pilot bearing in flywheel needs lubrication.
3. Release sleeve needs lubrication.
4. Misalignment.

RATTLES

1. Loose hub in clutch disk.
2. Worn release bearing.
3. Worn release part.
4. Worn pilot bearing.
5. Worn splines in hub or on shaft.
6. Worn driving pins in pressure plate.
7. Wear in transmission or drive line.
8. Worn transmission bearings.
9. Bent clutch shaft.
10. Unequal adjustment of release levers.
11. Misalignment.

GRABBING

1. Oil or grease on clutch disk facings.
2. Splined hub sticking on splined shaft.
3. Pressure plate sticking on driving studs.
4. Glazed or worn facings.
5. Sticking or binding release levers.
6. Sticking or binding clutch pedal or linkage.
7. Misalignment.

SLIPPING

1. Worn clutch disk facings.
2. Weak or broken springs.
3. Improper clutch or pedal adjustment.
4. Oil or grease on facings.
5. Warped disk.
6. Warped pressure plate.
7. Sticking release levers.
8. Pressure plate sticking or binding on studs.
9. Misalignment.

VIBRATIONS

1. Bent clutch shaft.
2. Defective clutch disk.
3. Dust in clutch.
4. Improper assembly of clutch to flywheel.
5. Use of rigid disk instead of flexible type.
6. Unmatched springs in pressure plate.
7. Misalignment of parts.

FAILURE

1. Clutch disk hub torn out.
2. Friction facings on clutch disk torn off or worn out.
3. Broken springs.
4. Incorrect adjustment of pressure plate.
5. Improper adjustment of clutch or pedal.
6. Splined disk hub stuck on splined shaft.

SUMMARY: TROUBLESHOOTING OF CLUTCHES

You can locate the basic problems using the charts here. Refer to your Technical Manual for specific problems that you might face in the servicing of a clutch.

TEST YOURSELF

QUESTIONS

1. Name four types of clutches.

2. What are the two kinds of disk cluches?

3. What characteristics must a clutch facing have? Name two.

4. How can you tell a rigid clutch disk from a flexible disk?

5. Fill in the blanks. "An overrunning clutch will _________ in one direction, but will_________ in the other."

6. True or False? "In an overcenter linkage, it is necessary to hold your foot on the pedal to keep the clutch disengaged."

7. True or False? "The clutch release bearing should rotate all the time in a plate clutch."

8. Grease or oil on a dry clutch disk surface usually causes:

1. Grabbing

2. Slipping

3. Both of the above

(Answers on page 9 in back of text.)

 Litho in U.S.A.

MECHANICAL TRANSMISSIONS / CHAPTER 3

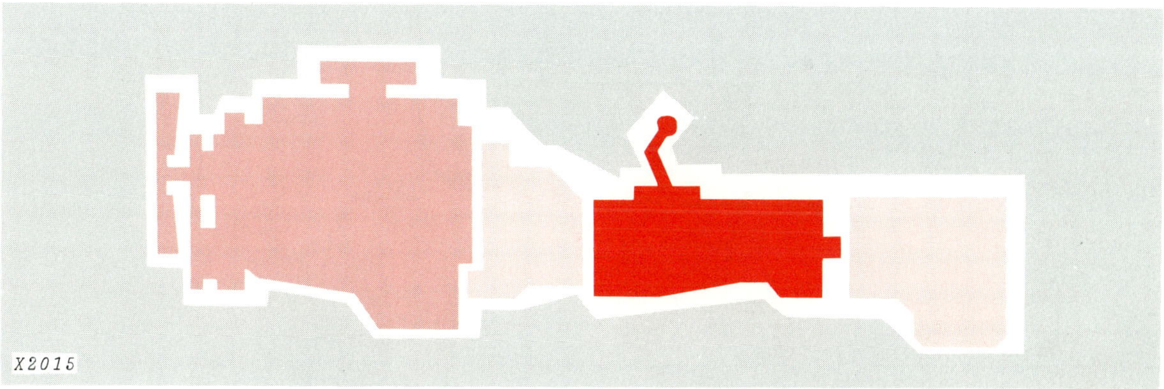

Fig. 1—Mechanical Transmission In A Complete Power Train

INTRODUCTION

A **mechanical transmission** is a train of gears that transfers and adapts the engine power to the drive wheels of the machine.

The transmission does two jobs:

1) Selects speed ratios for various travel speeds.

2) Reverses the travel of the machine.

The transmission can also power a special drive such as a PTO (Chapter 9).

Normally the transmission is located to the rear of the engine and clutch and in front of the differential or ring gear (Fig. 1).

TYPES OF MECHANICAL TRANSMISSIONS

Mechanical transmissions are of three major types:

- **Sliding Gear**
- **Collar Shift**
- **Synchromesh**

The SLIDING GEAR transmission has two or more shafts mounted in parallel or in line, with sliding spur gears arranged to mesh with each other and provide a change in speed or direction.

The COLLAR SHIFT transmission has parallel shafts with gears in constant mesh. Shifting is done by locking free-running gears to their shafts using sliding collars.

The SYNCHROMESH transmission also has gears in constant mesh. However, gears can be selected without clashing, by synchronizing the speeds of mating parts *before* they engage.

In all three cases, the transmission is shifted by hand or foot. So we call them *mechanical* transmissions.

Let's look at each type of transmission in detail.

SLIDING GEAR TRANSMISSIONS

The *sliding gear* transmission is a popular transmission for use in farm and industrial machines. Made up mainly of spur gears and shafts, it is simple but gives a variety of speeds.

There are two basic designs:

- **Input and output shafts parallel**
- **Input and output shafts in line**

Let's see how each one operates.

INPUT AND OUTPUT SHAFTS PARALLEL

The input shaft drives the output shaft, which transmits the power. There is normally a third shaft which reverses or further varies the power flow. All three shafts are in parallel and interact through the shifting of their gears. The transmission in Fig. 2 provides three forward speeds and one reverse speed.

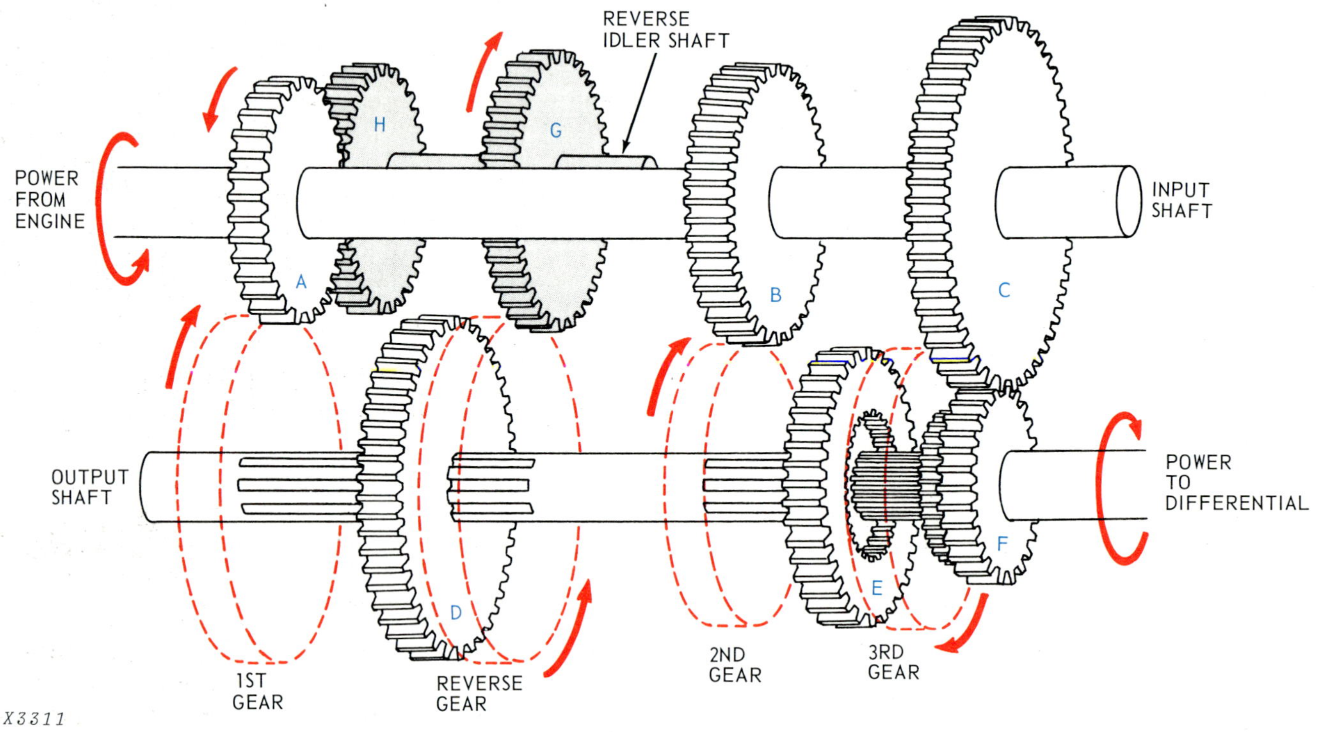

Fig. 2—Sliding Gear Transmission With Parallel Shafts

Gears D and E change the speed and direction (Fig. 2). Both of these gears slide on the output shaft, but are splined to it so as to turn it. All the other gears are fixed on their respective shafts, except gear F, which turns freely on the output shaft.

The dotted lines show how gears are shifted.

When gear D engages gear A, the transmission is in first or low gear.

When gear E engages gear B, the transmission is in second gear.

When gear E is engaged to gear F, the transmission is in third or high gear.

When gear D engages gear G, the rotation of the driven shaft is reversed. This places the idler gear shaft into the power train and feeds power in reverse. Gear G is powered by gear H, which is in constant mesh with gear A.

If neither gears D or E are engaged with another gear, the transmission is in neutral.

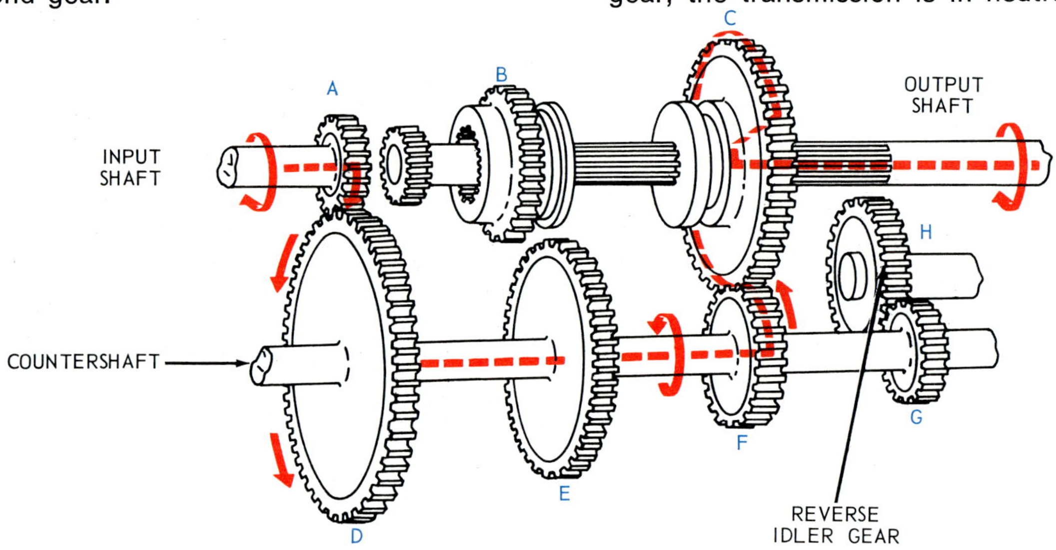

Fig. 3—Sliding Gear Transmission With Shafts in Line (First Gear Shown)

INPUT AND OUTPUT SHAFTS IN LINE

In this transmission (Fig. 3), input and output shafts lie in a straight line but are not connected. A third shaft, called a countershaft, "overlays" the two and transmits power between them.

The power flow in Fig. 3 is shown for low or first gear. Notice that power is flowing from gear A on the input shaft to gear D on the countershaft, which is also turning gear F. Gear F on the countershaft, in mesh with sliding gear C on the output shaft, causes that shaft to rotate and transmit power to the wheels.

A feature of this transmission is a high gear which bypasses the countershaft. It is provided by connecting gears A and B, and so locking the input and output shafts.

These two basic designs of sliding gear transmissions have many variations. They often provide as many as ten forward speeds and two reverse speeds.

Fig. 4—Sliding Gear Transmission In Crawler Tractor

COLLAR SHIFT TRANSMISSIONS

The *collar shift* transmission (Fig. 5) has parallel shafts with the gears in constant mesh. In neutral, the gears are free-running but when shifted they are locked to their shafts by sliding collars.

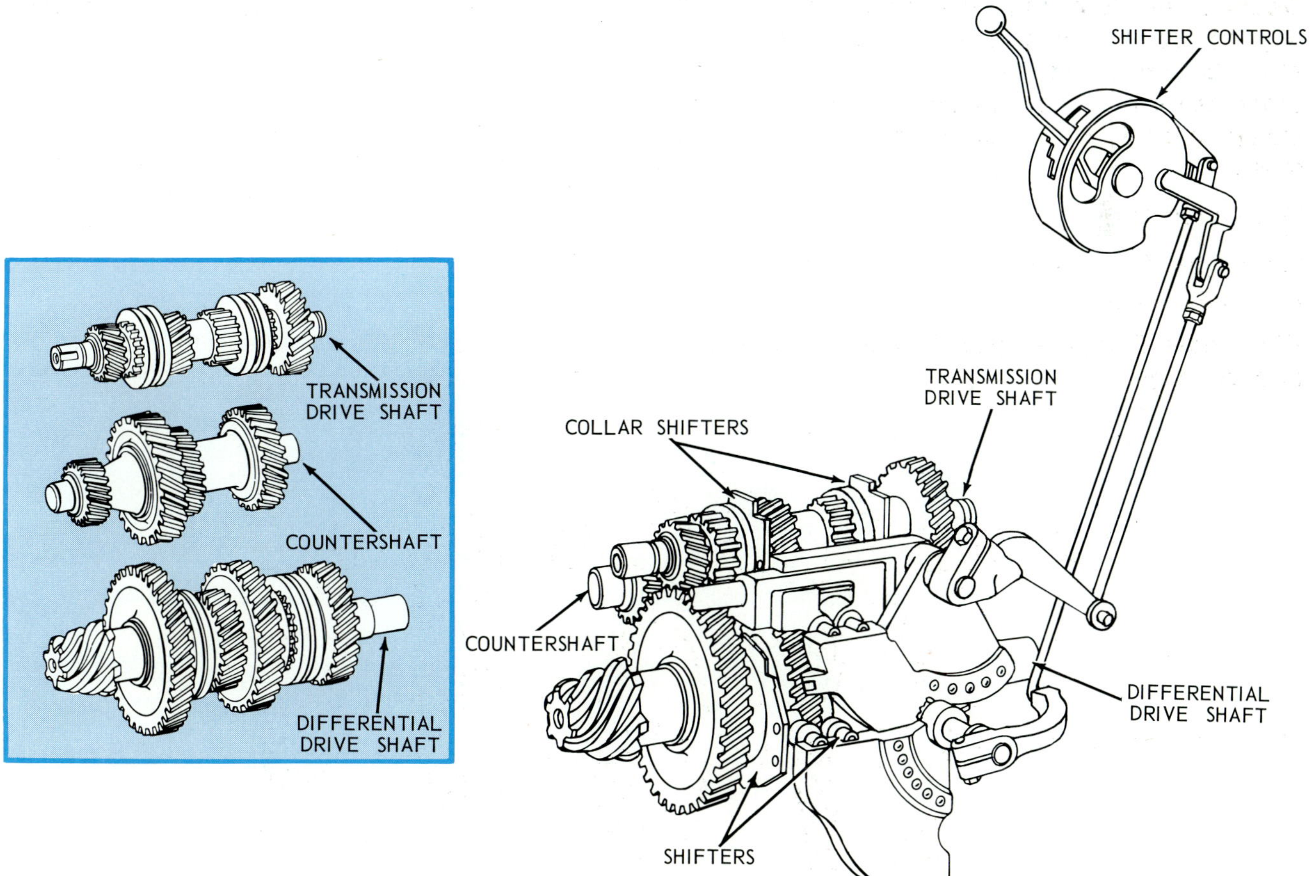

X2132

Fig. 5—Collar Shift Transmission For Wheel Tractor

 Litho in U.S.A.

The transmission shown in Fig. 5 has helical gears which remain in constant mesh. Helical gears usually provide quieter operation than the spur gears which the sliding gear transmission must use.

Since the gears are constantly in mesh and cannot slide on their shafts, one gear in each constant mesh gear set runs free on its shaft when in neutral.

When a gear is engaged, it is secured to its shaft by a collar or coupling device. The gear is disengaged when the collar is released and the gear turns freely again.

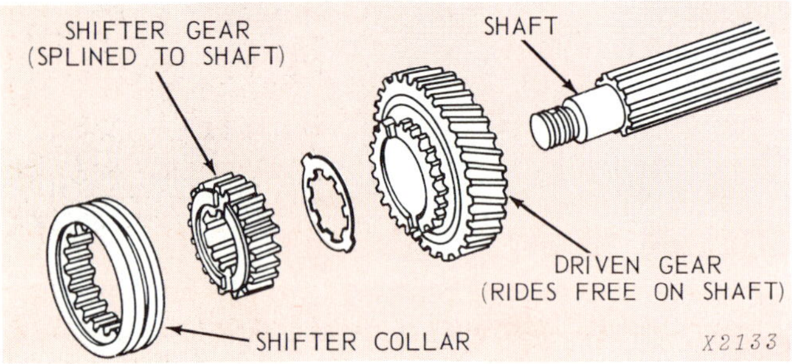

Fig. 6—Typical Shifter Collar And Mating Gear

Let's take a close look at the collar which locks a gear to the shaft. In Fig. 6, we see a typical shifter collar and its mating gear. The side of the gear will usually carry short splines. The shifter collar with internal splines is splined to a shifter gear which is, in turn, splined to the shaft. (Sometimes the shifter collar is splined directly to the shaft without a shifter gear).

Power is transmitted by shifting the collar so its internal splines are engaged with the external splines on the driven gear and the shifter gear.

The splines on the gears and collar have rounded ends for easier shifting.

OPERATION

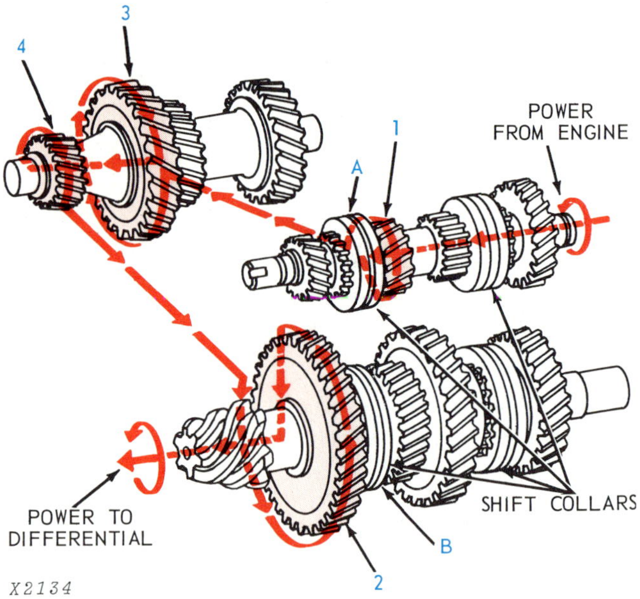

Fig. 7—Collar Shift Transmission
(Power Flow In First Gear Shown)

Now let's look at a collar shift transmission in operation. Fig. 7 shows a transmission which provides eight forward speeds and four reverse speeds. (The complete unit is shown in Fig. 5).

Power flow is indicated for first gear. Notice that collars A and B are locking gears 1 and 2 to their respective shafts, transmitting power to the differential and the drive wheels. Power flows through the mating gears 3 and 4 as shown by the arrows.

All eight speeds are obtained by securing the different gears to the shafts, using the collars.

Fig. 8—Collar Shift Transmission In Wheel Tractor

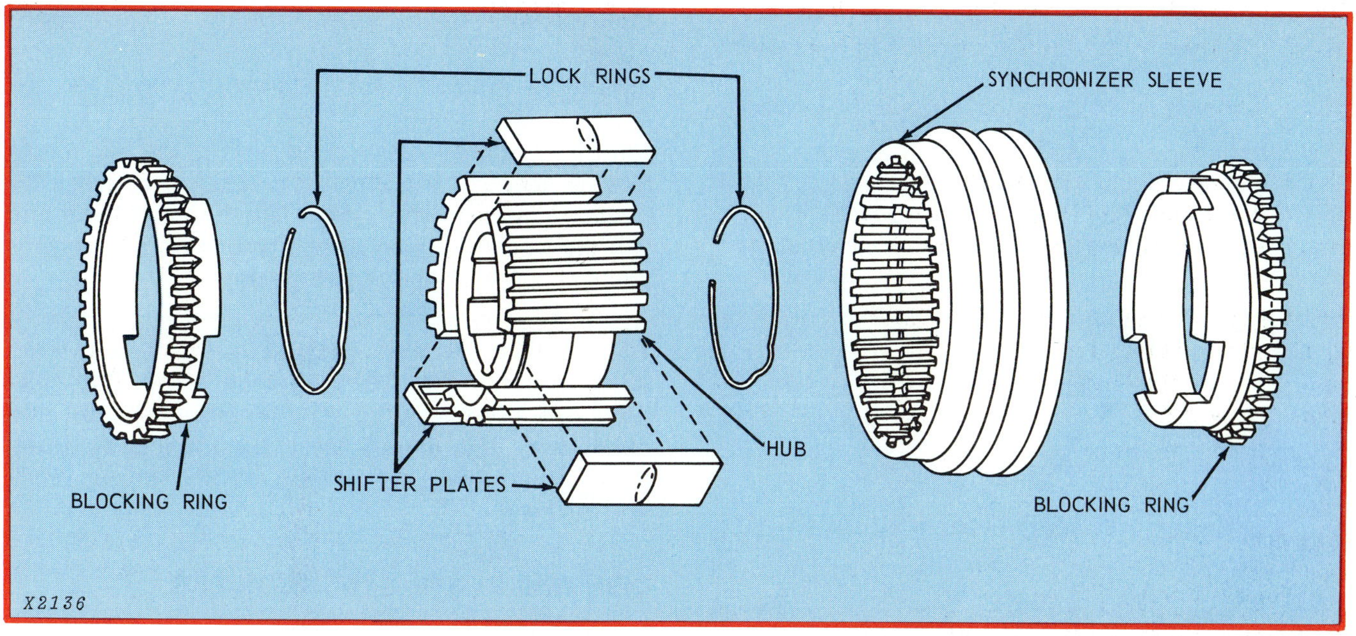

Fig. 9—Exploded View Of Block-Type Synchronizer

SYNCHROMESH TRANSMISSIONS

The *synchromesh* transmission is basically a constant mesh, collar-shift transmission with an extra device (called a synchronizer) to equalize the speed of the mating parts *before* they engage.

The synchronizer is used in all manual automotive transmissions and is common in other machines where shifting while moving is required.

Four common synchronizers are:

- **Block Synchronizer**
- **Disk-And-Plate Synchronizer**
- **Plain Synchronizer**
- **Pin Synchronizer**

Before we look at each of these synchronizers, remember that they have a common function—to help in engaging gears by synchronizing the speeds of the gear and the shaft *before* the connection is made. All use friction to synchronize the mating parts.

BLOCK SYNCHRONIZER

The parts of a block synchronizer are: the blocking rings, lock rings, hub, synchronizer sleeve, and shifter plates (Fig. 9).

The hub is splined to the shaft and the synchronizer sleeve is mounted on the hub. The two lock rings hold the shifter plates out against the sleeve. The bronze blocking rings each have three slots into which the ends of the shifter plates fit.

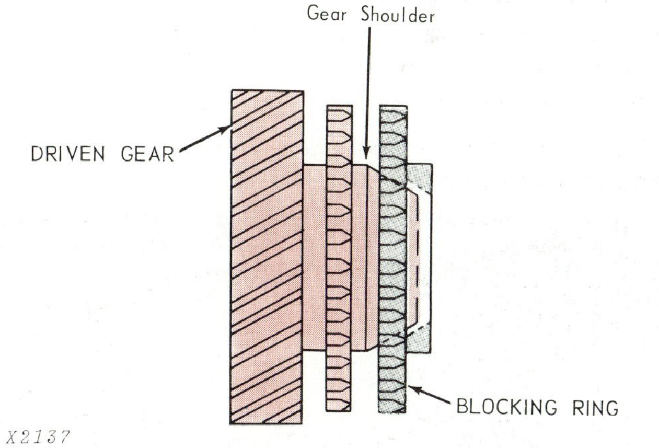

Fig. 10—Matching Cone Areas Of Gear And Blocking Ring

The inner surfaces of the blocking rings are cone-shaped and match the conical shape of the driven gear shoulders which they contact (Fig. 10). These cone-shaped surfaces provide the frictional force to synchronize the speed of the main shaft and the gear.

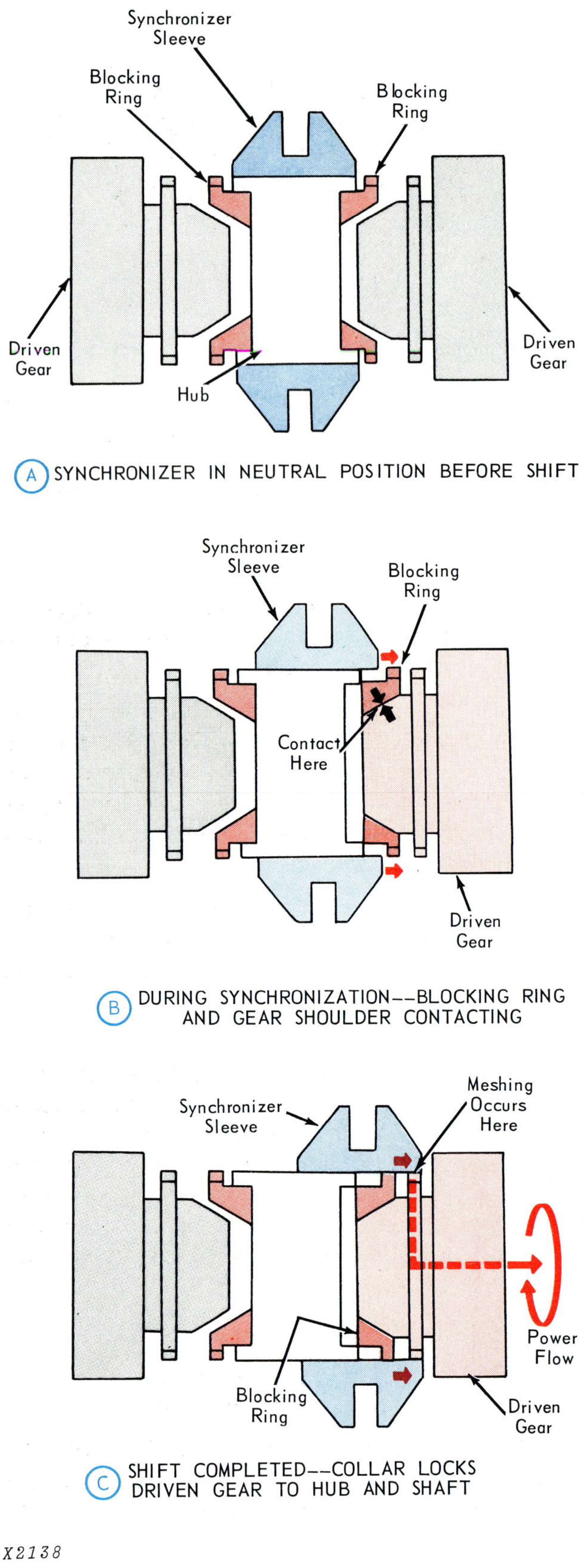

Fig. 11—Block Synchronizer In Operation

Operation Of Block Synchronizer

In neutral, the parts are as shown in Fig. 11A.

During synchronization (Fig. 11B), the sleeve is moved toward the selected gear, pushing the blocking ring to the right. The ring contacts the shoulder of the driven gear and begins to synchronize the speeds of the two parts.

To complete the shift (Fig. 11C), the sleeve teeth pass through the blocking ring teeth (because both are now rotating at the same speed) and mesh with the clutch teeth on the driven gear. Power then flows as shown by the dotted lines and arrows.

DISK-AND-PLATE SYNCHRONIZER

Shown in Fig. 12 are the parts which make up the disk-and-plate synchronizer.

Note the relationship of the parts to each other.

This synchronizer uses friction disks and plates to bring both gears to the same speed for engaging.

The blocker (B) rides on and is driven by the synchronizer gear (A). The blocker is held in neutral on the gear by four detent balls (C) and springs (D).

The synchronizer drum (G) is driven by the output gear (H). Therefore, the drum is traveling at a different speed than the blocker.

To bring the blocker and the drum to the same speed, three synchronizer disks (E) and two separator plates (F) are housed between the drum and blocker.

The tangs on the synchronizer disks locate themselves inside the synchronizer drum slots and so travel at the same speed as the drum and gear (H).

The separator plates have internal tangs which hold them in position on the blocker, giving them the same rotating speed as the blocker and synchronizer gear.

 Litho in U.S.A.

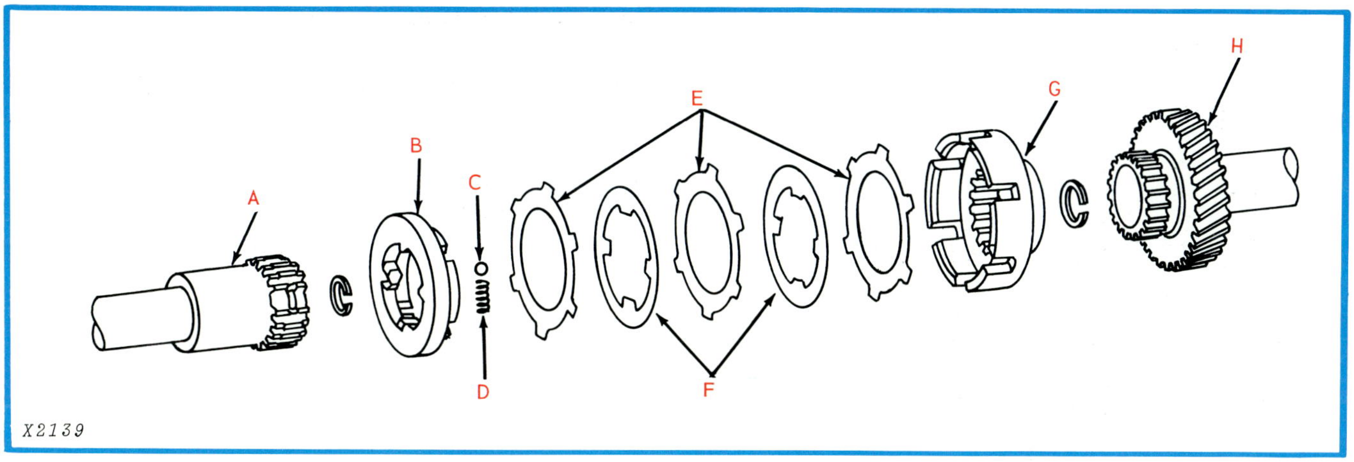

A—Input Shaft With Synchronizer Gear
B—Blocker
C—Detent Ball
D—Detent Spring
E—Synchronizer Disk
F—Separator Plate
G—Synchronizer Drum
H—Output Gear And Shaft

Fig. 12—Exploded View Of Disk-And-Plate Synchronizer

Operation

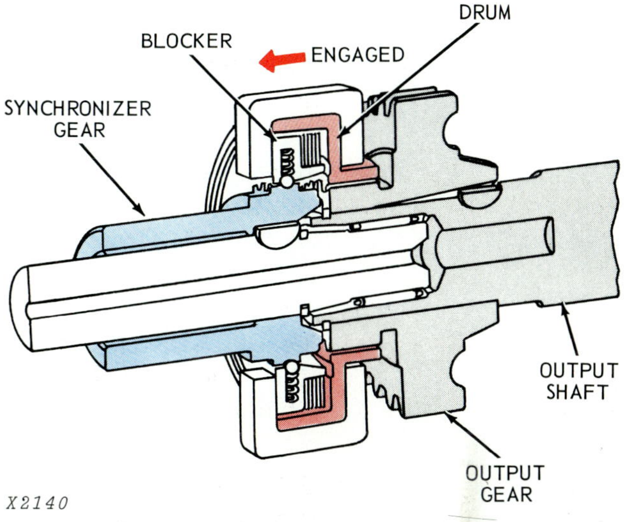

Fig. 13—Operation Of Disk Synchronizer

Shifts are made by coupling the output gear to the synchronizer gear (Fig. 13).

When the shifter fork moves the drum forward, the synchronizer disks (which are connected to the drum) and the separator plates (which are connected to the blocker) make contact as shown. As the contact is made, the blocker tends to rotate about its gear instead of moving forward in the direction the shifter fork is moving. This action forces the drive lugs on the blocker to engage into the outer edge of the detent ball depression cut into the synchronizer gear.

Since the blocker is now locked into the synchronizer gear, extra forward motion of the shifter tends to compress the disks and plates into proper clutching action. This action matches the speed of the synchronizer gear and blocker to the speed of the output gear.

As soon as both gears are moving at the same speed, the thrust force locking the blocker into the synchronizer gear is relieved and the blocker backs out of its locked position with the aid of the detent balls and springs.

This synchronizing action is almost instantaneous. When both gears are moving at the same speed, the drum can then be moved forward to engage both gears. A ridge on the highest teeth of the synchronizer gear limits the forward travel of the drum. This prevents the synchronizer drum from traveling too far forward and losing its contact with the output gear.

PLAIN SYNCHRONIZER

The plain-type synchronizer is similar to the block-type synchronizer, but has fewer parts.

The hub is splined to the shaft (Fig. 14). The sleeve is mounted on the hub and connected to it by the springs and balls located radially around the hub.

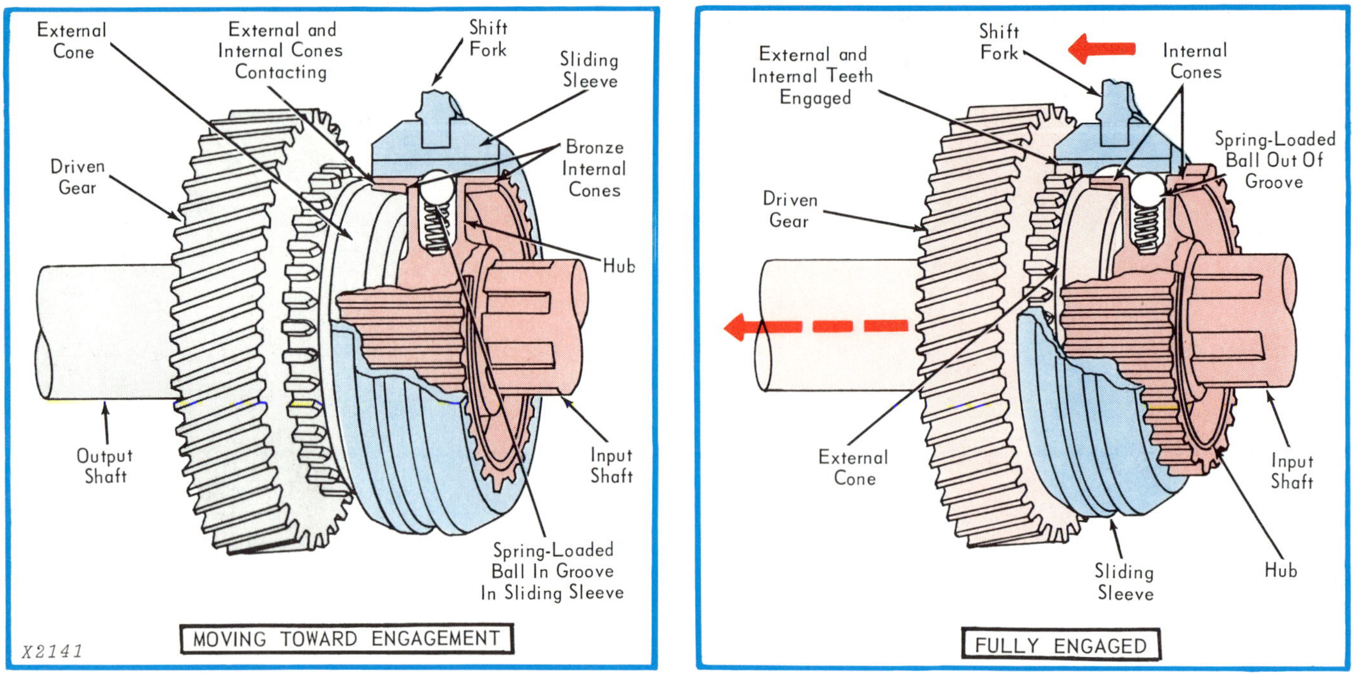

Fig. 14—Plain Synchronizer

Operation

To engage, the assembly is moved toward the gear to be engaged as shown at left in Fig. 14. The cone-shaped friction area of the hub contacts the cone-shaped area of the gear and accelerates or decelerates it to the same speed as the synchronizer. The small coil springs hold the balls in the groove in the center of the sleeve. This prevents the sleeve from moving off the hub and the sleeve teeth from engaging with the gear teeth until enough pressure has been put on the cone-shaped friction area to synchronize the two parts. Then further movement of the sleeve pushes the balls in, compressing the springs, and the sleeve teeth engages the teeth on the gear, completing the shift. This action is shown at right in Fig. 14.

PIN SYNCHRONIZER

Fig. 15 illustrates a pin-type synchronizer. The actuating hub is splined to the shaft and rotates with it. The outer stop rings are splined to the ends of the gears. The stop ring and pin assemblies are loosely pinned to the actuating hub.

Operation

When the actuating hub is moved to either the right or left, the stop ring and pin assemblies hold the loose-fitting pin against the side of the holes in the actuating hub. The actuating hub is prevented from engaging the gear by the chamfered shoulder on the stop ring and the pin assembly.

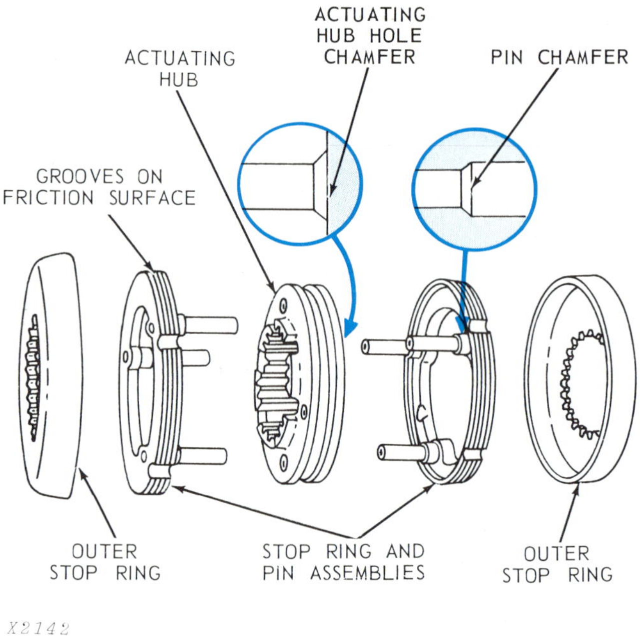

Fig. 15—Pin Synchronizer

When all parts are rotating at the same speed, the force between the pin and actuating hub is reduced. The hub can then move over the large base of the pins, and the internal splines on the hub can engage the splines on the gear. Slight chamfers on the pin and actuating hub, as well as rounded ends of the splines on the hub and gear, permit these parts to align themselves and mesh easily.

 Litho in U.S.A.

SYNCHROMESH TRANSMISSION IN OPERATION

Now that we've explained the different types of synchronizers, let's see how they work in an actual synchromesh transmission.

Fig. 16 shows a three-speed transmission which is synchronized in second and third gear. Notice that this transmission uses both a sliding gear, a synchronizer to change speeds, and in-line shafts.

Gear (A) is splined to the input shaft and is in constant mesh with gear (D). Gear (B) turns freely on the output shaft.

Low and reverse is provided by sliding gear (C) into mesh with gear (F) or gear (G), the reverse idler.

Second gear is provided by sliding the synchronizer towards gear (B), synchronizing it and locking it to the output shaft. Power flow in Fig. 16 is for second gear.

Third or high gear is provided by sliding the synchronizer towards gear (A), synchronizing it, and locking the input shaft to the output shaft. Power then flows directly through the transmission from input to output.

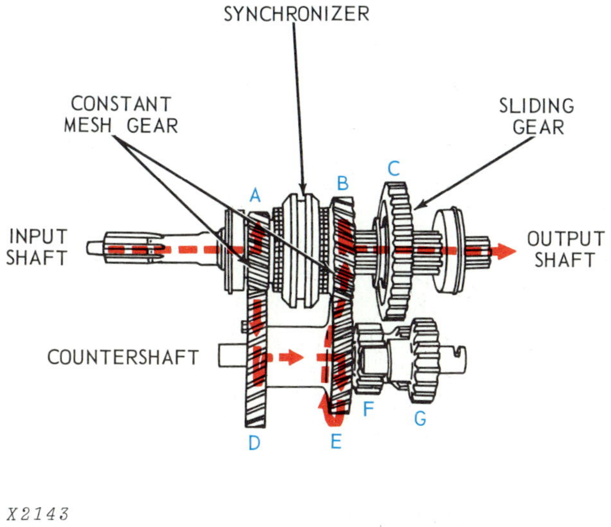

Fig. 16—Three-Speed Synchromesh Transmission (Power Flow For Second Gear Shown)

A complete synchromesh transmission for a wheel tractor is shown in Fig. 17.

Fig. 17—Synchromesh Transmission In a Wheel Tractor

SHIFT CONTROLS

Now let's look at the *shift controls* which actuate the various mechanical transmissions.

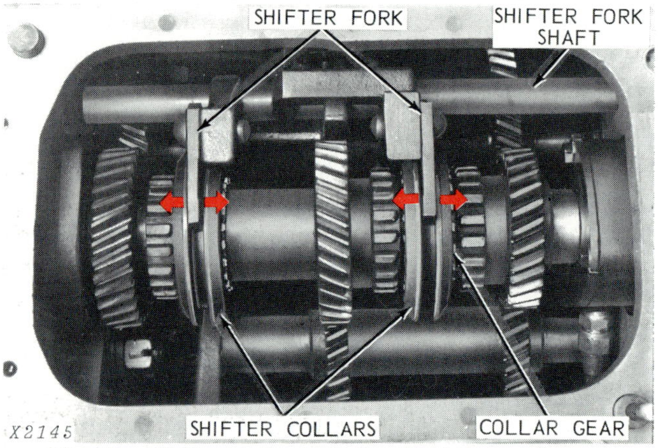

Fig. 18—Shifter Forks In Collar Shift Transmission

In sliding gear, collar shift, and synchromesh transmissions, the gears or collars are moved by means of a shifter fork which fits in a groove in the collar or gear (Fig. 18). This fork is connected to and moved manually by the gearshift lever. How this connection is made varies.

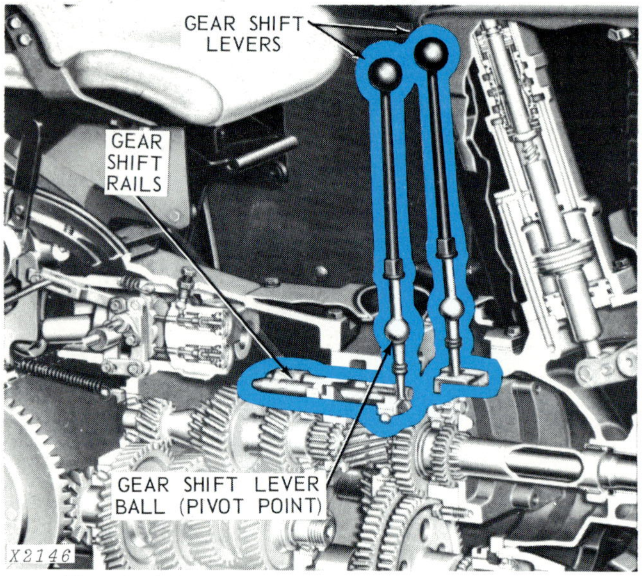

Fig. 19—Direct Transmission Shifting Mechanism

Let's briefly look at two popular shifter mechanisms:

1) Direct-Type Shifters

2) Cam-Type Shifters

DIRECT SHIFTERS

In the direct shifter, the gearshift lever reaches directly into the transmission from the operator's platform (Fig. 19).

The shifter forks, which slide the gears or collars, are attached to the gearshift rails or shafts. The gearshift lever in Fig. 19 has a ball that fits into a socket in the top of the transmission cover. This permits swiveling of the gearshift lever into the various shift positions. Movement of the lever causes the finger on the lower end to select and move one of the rails, its forks, and a gear or collar.

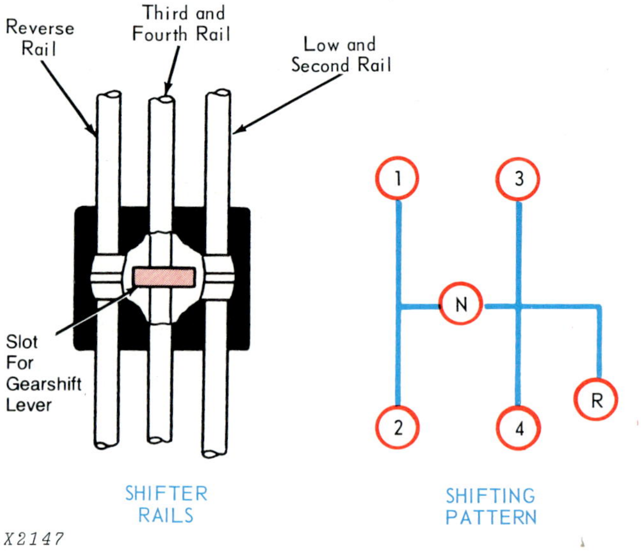

Fig. 20—Shift Pattern On The Four-Speed Transmission Shown in Fig. 19.

For example, in a four-speed transmission (Fig. 20), as the gearshift lever is moved from neutral (N) to the left, the finger on the lower end of the lever moves into the slot in the low-and-second speed rail. This action selects the rail, forks, and gears to be moved. Then, as the lever is pushed forward into low speed, the rail assembly is pushed toward the rear, engaging the low gear.

Litho in U.S.A.

Locking Of Direct Shifters

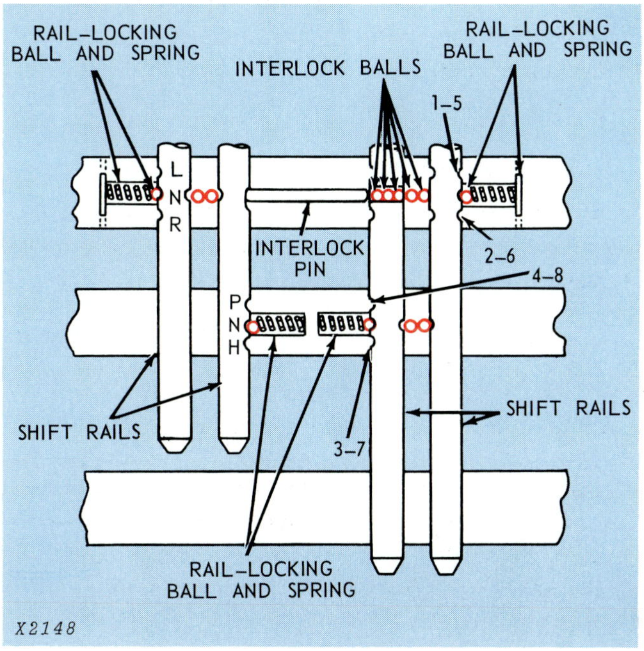

Fig. 21—Locking System For Shifter Rail

To hold the shift rail in position and to prevent other rails from moving, a locking system must be used. Fig. 21 illustrates a popular type.

The shift rails contain spring-loaded *detent balls* that drop down into undercuts on the rails as the rails are moved into the various shift positions. This locks the gears in mesh.

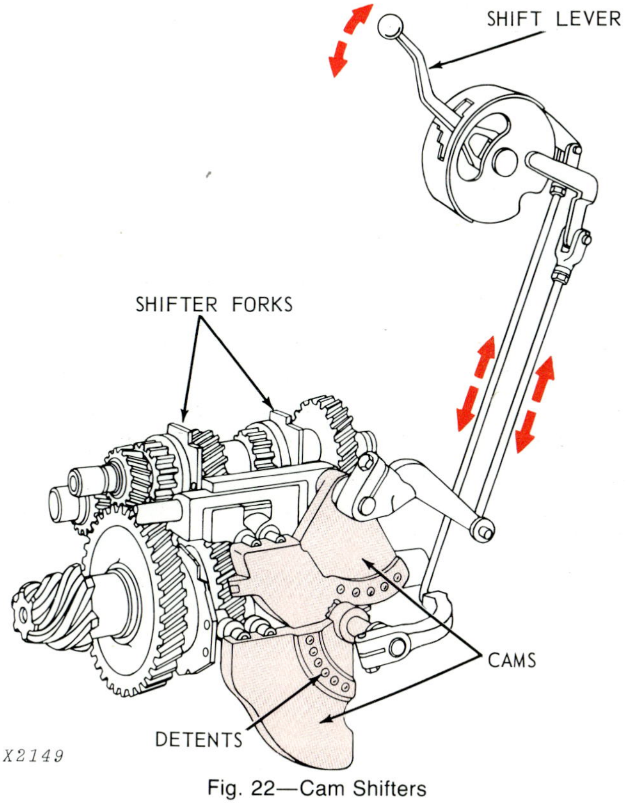

Fig. 22—Cam Shifters

To prevent movement of more than one rail at a time, *interlock balls* are placed *between* the shift rails as shown. When a rail is moved, the balls are forced out of the undercuts in the side of the rail and into the undercuts on the other rails, locking them firmly into position.

CAM SHIFTERS

Fig. 22 shows a cam shifter with a shifter cam for the range gearing and one for the speed gearing. The cams are operated through a linkage of rods and arms. When the shift lever is moved, the linkage moves the shifter cam about its pivot point. The cam in turn moves the shifter forks.

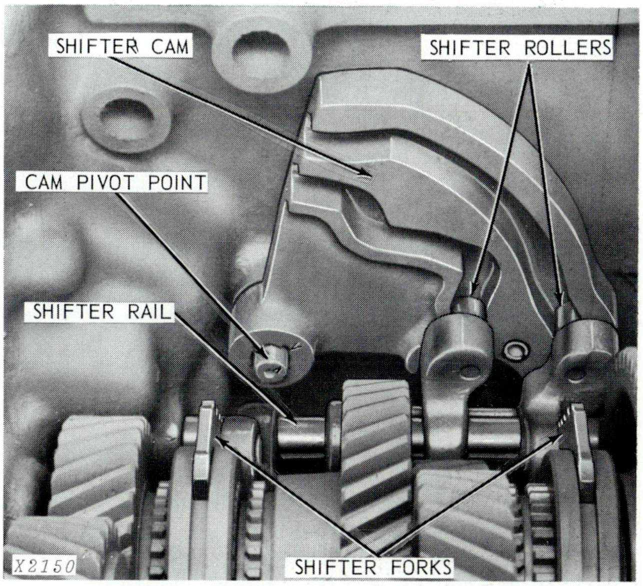

Fig. 23—Shifter Cam And Rail Assembly

The shifter cam (Fig. 23) is a plate with two irregular grooves. Rollers on the rail and fork assembly slide in these grooves. When the cam pivots, the rollers, sliding in the irregular grooves, move the forks to engage the selected gear.

Like the direct shifter, this unit has detents which lock the transmission in the designed gear. These detents are located in the cam itself as shown in Fig. 22.

ADJUSTMENTS

When a transmission is disassembled to replace bearings or gears, the shafts may require preload or endplay adjustments as explained in Chapter 1. The specific procedures and measurements are supplied by the manufacturer in his technical manual.

GENERAL MAINTENANCE

Mechanical transmissions need little maintenance other than periodic oil changes. However, when you repair a transmission, examine the whole gear train to locate worn or faulty parts and repair or replace them at that time. Thus you may prevent a breakdown and the need to disassemble the transmission once again.

Some things to look for:

1. Excessive gear tooth wear or broken teeth.

2. Worn-out bearings.

3. Broken or distorted shifter detent springs or scored or flat detent balls.

4. Damaged or plugged transmission oil lines or passages in shafts.

5. In sliding gear transmissions, badly worn splines.

6. In collar shift transmissions, excessively worn teeth in shifter collars and hubs.

7. In synchromesh transmissions, scored contact surfaces on blockers, drums, or disks.

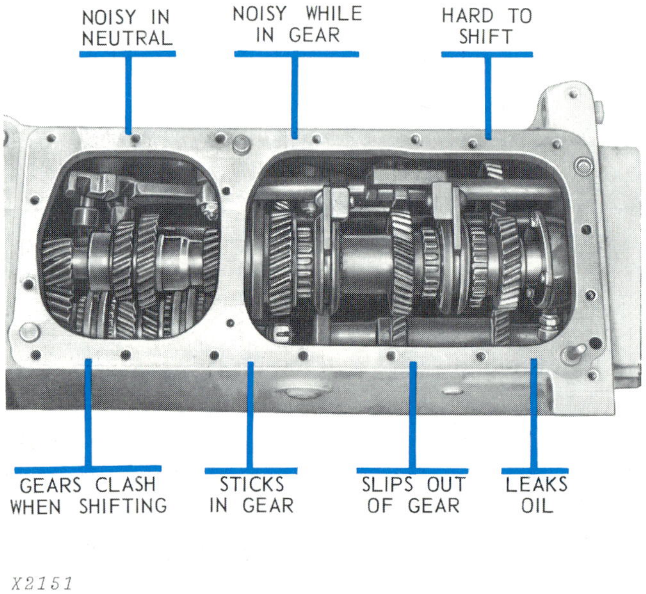

Fig. 24—Troubleshooting Problems With Mechanical Transmissions

TROUBLESHOOTING

These troubleshooting charts are given as a general guide to the common transmission failures. They list what the cause might be, and the remedy.

TRANSMISSION NOISY IN NEUTRAL

Possible Cause	Remedy
1. Transmission not aligned with engine.	1. Align.
2. Bearings dry, badly worn, or broken.	2. Lubricate or replace.
3. Transmission oil level low.	3. Refill.
4. Gears worn, scuffed, or broken.	4. Replace.
5. Countershaft sprung or badly worn.	5. Replace.
6. Excessive endplay of countershaft.	6. Adjust or replace worn parts.

TRANSMISSION NOISY WHILE IN GEAR

1. All causes noted in "Transmission Noisy In Neutral."	1. Same as above.
2. Main shaft rear bearing worn or broken.	2. Replace.
3. Gear teeth worn.	3. Replace gear.
4. Engine vibration dampener defective.	4. Replace or adjust.
5. Speedometer drive gears worn.	5. Replace.
6. Clutch friction disk defective.	6. Replace.
7. Gears loose on main shaft.	7. Replace worn parts.
8. Transmission oil level low.	8. Refill.

TRANSMISSION HARD TO SHIFT

1. Engine clutch not releasing.	1. Adjust.
2. Sliding gear tight on shaft splines.	2. Clean splines or replace shaft or gear.
3. Shift linkage out of adjustment.	3. Adjust.
4. Main shaft splines distorted.	4. Replace or clean splines.
5. Synchronizing unit damaged.	5. Replace defective parts.
6. Sliding gear teeth damaged.	6. Replace.

GEARS CLASH WHEN SHIFTING

1. Clutch not releasing.	1. Adjust.
2. Synchronizer unit defective.	2. Replace defective parts.
3. Gears sticky on main shaft.	3. Free up gears. Replace defective parts.

TRANSMISSION STICKS IN GEAR

1. Clutch not releasing.	1. Adjust.
2. Detent balls stuck.	2. Free.
3. Shift linkage out of adjustment not lubricated.	3. Adjust or lubricate.
4. Sliding gears tight on shaft splines.	4. Clean splines or replace shaft or gears.

TRANSMISSION SLIPS OUT OF GEAR

1. Shift linkage out of adjustment.	1. Adjust.
2. Gear loose on shaft.	2. Replace shaft or gear.
3. Gear teeth worn.	3. Replace gear.
4. Excessive endplay in gears.	4. Replace worn or loose parts.
5. Lack of spring tension on shift lever detent.	5. Install new spring.
6. Badly worn transmission bearings.	6. Replace.

TRANSMISSION LEAKS OIL

1. Oil level too high.	1. Drain to proper level.
2. Gaskets damaged or missing.	2. Install new gaskets.
3. Oil seals damaged or improperly installed.	3. Install new oil seals.
4. Oil throw rings damaged, improperly installed, or missing.	4. Install oil throw rings properly.
5. Drain plug loose.	5. Tighten.
6. Transmission case bolts loose, missing, or threads stripped.	6. Tighten or replace.
7. Transmission case cracked.	7. Replace.
8. Lubricant foaming excessively.	8. Use recommended lubricant.

TEST YOURSELF

QUESTIONS

1. What are the three common types of mechanical transmissions?

2. (True or False?) "In some sliding gear transmissions, the input and output shafts are locked together in high gear."

3. What is one advantage of the collar shift transmission over the sliding gear?

4. Which type of transmission allows "on-the-go" shifting while the machine's wheels are still rolling?

5. What is the purpose of the countershaft?

6. Does a collar shift transmission normally have spur or helical gear teeth?

7. What is a chamfer?

(Answers on page 9 in back of text.)

 Litho in U.S.A.

HYDRAULIC ASSIST TRANSMISSIONS / CHAPTER 4

Fig. 1—Hydraulic Assist Transmission In A Complete Power Train

INTRODUCTION

The **hydraulic assist transmission** is a train of gears which can be shifted without interrupting the flow of power.

The gears are kept in constant mesh while two or more hydraulic clutches control the flow of power "on the go."

The parts work together as follows:

- **Hydraulic clutches—control the power flow**
- **Gear train—transmits the power flow**

When the operator shifts gears, hydraulic oil engages the clutches that route power to the selected gears.

A *hydraulic clutch* is normally an alternating pack of friction disks and plates (Fig. 2).

The clutch is *engaged* when pressure oil is sent to push the piston against the disks and plates, clamping them together. The disks are splined to the drum, while the plates are splined to the hub. As a result, input power through the drum is sent on through the hub to the output by the engaged clutch.

The clutch is *disengaged* when oil pressure is released and the piston moves away from the clutch pack. This frees the disks from the plates and the power flow is stopped. Spring action or oil pressure on the other side of the piston may be used to help release the disks and plates.

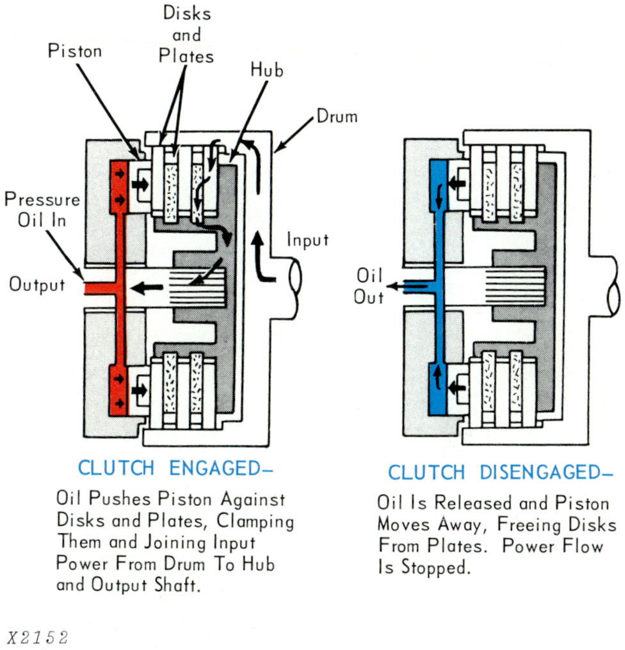

Fig. 2—Hydraulic Clutches Control The Power Flow In Hydraulic Assist Transmissions

The *gear train* usually takes one of two forms:

- **Countershaft**
- **Planetary**

Let's look at each one in detail.

Litho in U.S.A.

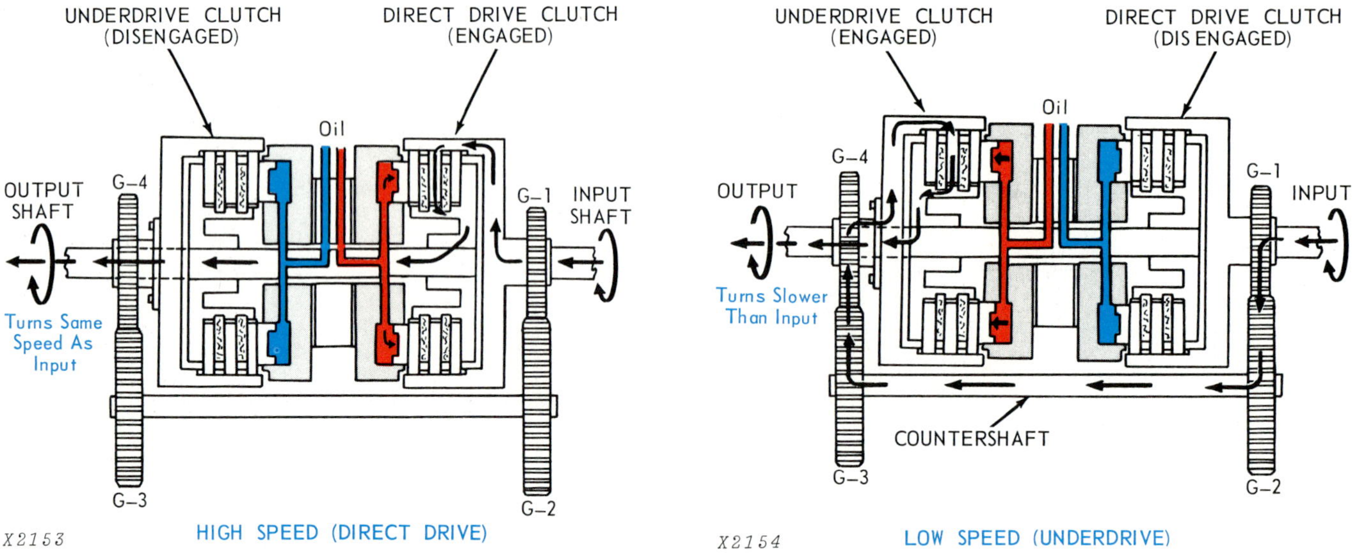

Fig. 3—Countershaft Transmission (High-Low Speed Unit)

COUNTERSHAFT TRANSMISSIONS

Countershaft transmissions allow one set of gears to be shifted without disturbing the other gear ratios in the transmission.

Let's look at three common models—high-low speed units, reverser units, and high-low-reverse units.

HIGH-LOW SPEED UNIT

The simplest countershaft transmission is the high-low speed unit shown in Fig. 3. It has two hydraulic clutches and four gears in mesh. Whenever one clutch is engaged, the other is disengaged.

The input and output shafts are not connected unless the direct drive clutch is engaged.

When engaged, this clutch sends input power from the outer drum through to the inner hub as explained in Fig. 2.

In direct drive, both shafts are locked together and rotate at the same speed. This gives *high speed* as shown in Fig. 3.

For *low speed,* the underdrive clutch is engaged. Gear G-1 is fixed on the input shaft so power is routed through mating gear G-2, along the countershaft, out through gears G-3 and G-4 and to the underdrive clutch drum. Power then flows through the engaged clutch pack to the inner hub which is fixed to the output shaft.

The output shaft turns at a slower speed than the input shaft, giving the low speed. The actual speed depends upon the ratio of the two gear sets.

Underdrive means that the ouput shaft turns *slower* than the input. In some units the output turns *faster* than the input and these are called **overdrives.**

REVERSER UNIT

To get a reverse speed from the countershaft unit, another gear (G-5) is added between gears G-1 and G-2 as shown in Fig. 4.

The countershaft will now rotate in the same direction as the input shaft.

In **forward speed,** the forward clutch is engaged and power flows straight through, bypassing the countershaft.

In **reverse speed,** the reverse clutch is engaged and locks gear G-4 to the output shaft. Power is transmitted through the countershaft as shown by the arrows, causing the output shaft to rotate in the opposite direction from the input shaft.

An actual reverser unit is shown (outlined in blue) in Fig. 5.

The engine power source is shown at right, while the transmission gear train is shown at left center.

The reverser unit gives on-the-go reversing, while the transmission gear train sets up the basic gear ratios. (The gear train is a sliding gear type —see Chapter 3.)

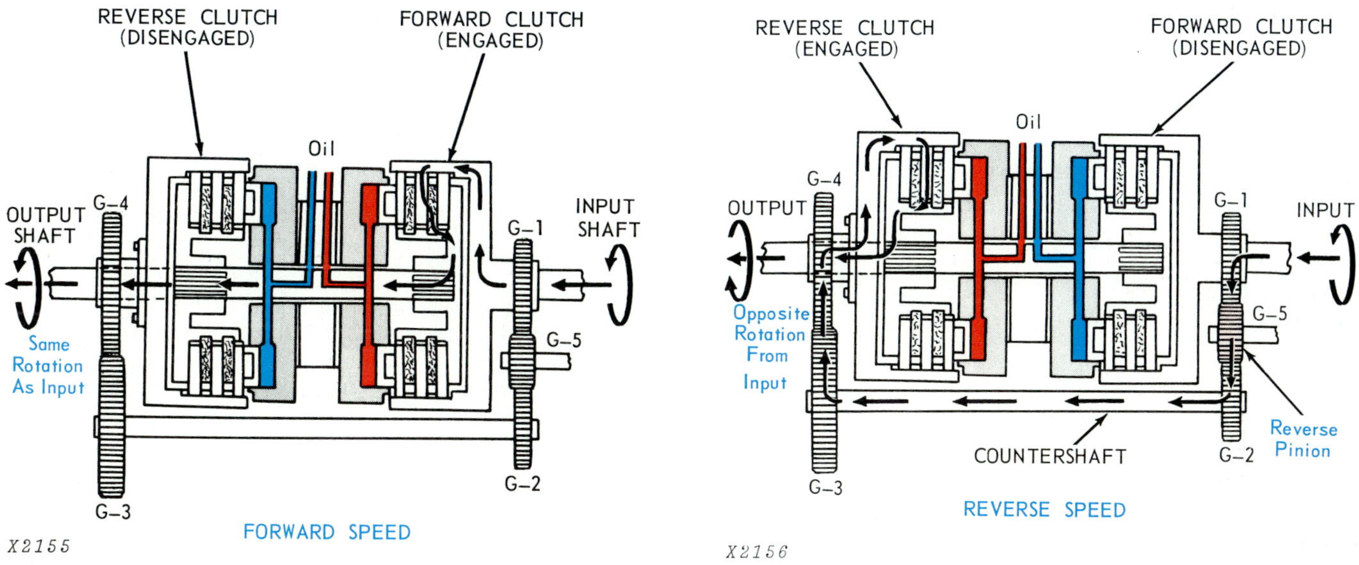

Fig. 4—Countershaft Transmission (Reverser Unit)

HIGH-LOW-REVERSE UNIT

If both a reverse and an underdrive countershaft are used in the same transmission with three clutches, these options are available:

1) High (direct drive)

2) Low (underdrive)

3) Reverse

Extra speeds can be obtained by simply adding more countershafts and more clutches.

PLANETARY TRANSMISSIONS

Planetary gears work like our solar system. The pinions or planet gears each turn on their own axis and, at the same time, rotate around the sun gear (Fig. 6). This is much like the earth and other planets rotating around the sun.

When used with hydraulic clutches and brakes, the planetary can serve as a direct coupler, a reduction gear, or a reversing gear.

A planetary gear set allows gear ratios to be changed without actually engaging or disengaging gears. As a result, there is little or no interruption of the power flow.

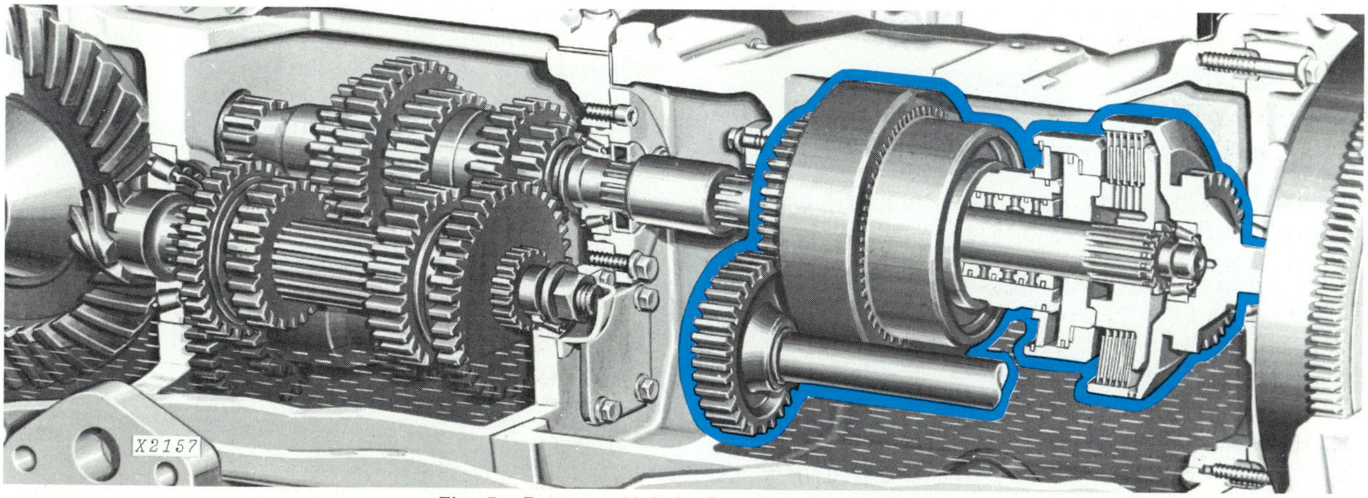

Fig. 5—Reverser Unit In Complete Power Train

Planetaries are compact units which allow the input and output shafts to rotate on the same axis and eliminate the countershaft.

In planetary sets, the gear load is spread over several gears, decreasing the load on each tooth.

The planetary system also spreads the load evenly around the circumference of the system, eliminating the sideways stress on the shafts.

Operation of basic planetaries is explained in Chapter 1.

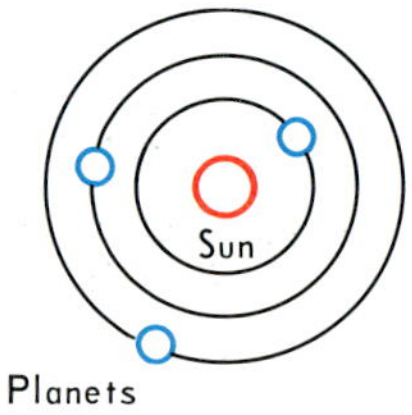

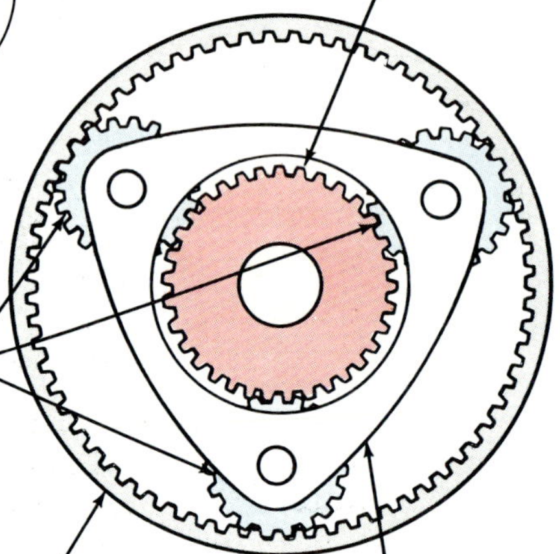

Fig. 6—Parts Of The Planetary System

The basic parts of the planetary system are shown in Fig. 6. They are:

- **Sun gear—center of system**
- **Planet pinions and carrier—rotate around sun gear**
- **Ring gear—surrounds the other parts**

A planetary transmission uses hydraulic clutches and brakes to control the rotation of the planetary parts.

The planetaries may be simple or compound units as we'll see now.

SIMPLE PLANETARY TRANSMISSIONS

The two-speed planetary shown in Fig. 7 has one clutch and one brake. The clutch controls the direct drive or high speed, while the brake controls the underdrive or low speed.

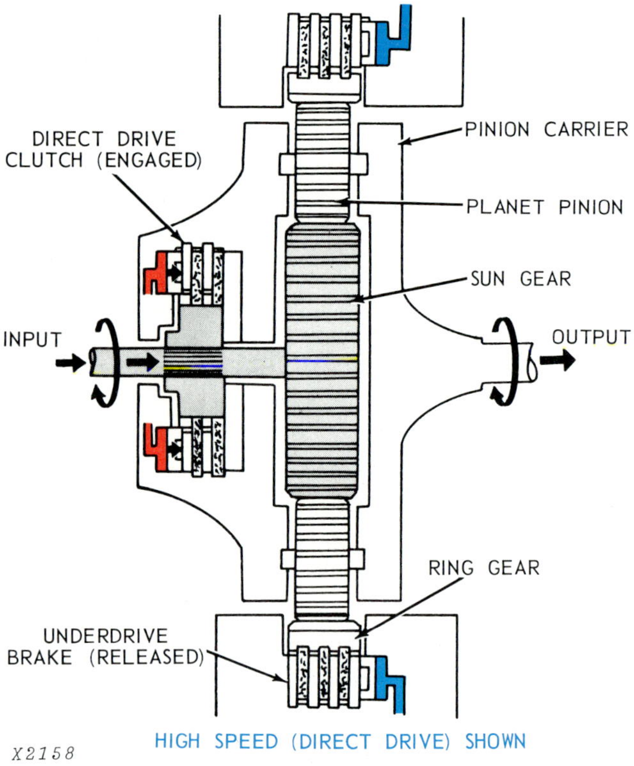

Fig. 7—Simple Planetary System (High-Low Speed Unit)

Planetaries work on the principle that **when any two parts are locked together, all three parts rotate as a unit.**

In Fig. 7, the sun gear shaft is the power input, while the output shaft is fastened to the planet pinion carrier.

For **high speed** (shown), the direct drive clutch engages and locks the sun gear to the planet pinion carrier, making the input and output shafts rotate in unison for a direct drive.

For **low speed,** the direct drive clutch is released while the underdrive brake is engaged, holding the ring gear stationary. The rotating sun gear then turns the planet pinions against the stationary ring gear and forces the pinion carrier to rotate as the planet pinions "walk" around the ring gear. As a result, the pinion carrier rotates in the same direction as the sun gear, *but at a reduced speed.* So the output shaft turns slower than the input shaft.

For **neutral,** both the clutch and brake are disengaged and the output shaft stops turning.

The unit shown in Fig. 7 is part of a shift-on-the-go transmission having high-low speed options.

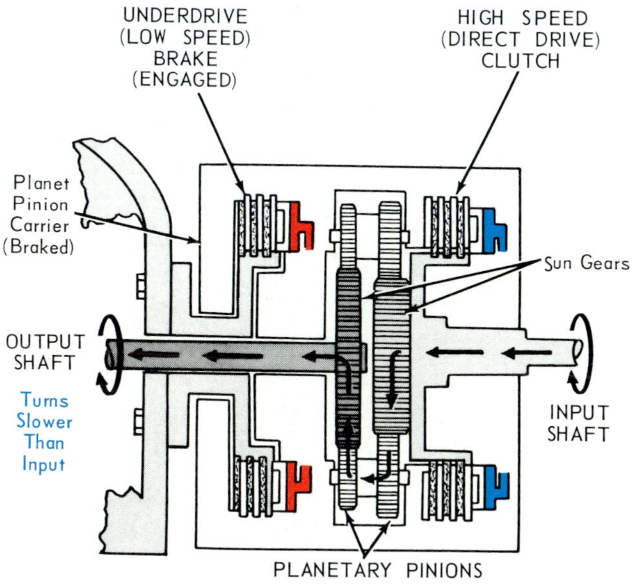

X2159 LOW SPEED SHOWN

Fig. 8—Simple Planetary System (High-Low Unit)

Fig. 8 shows a two-speed planetary with no ring gear.

For **low speed** (shown), the underdrive brake is engaged, stopping the planet pinion carrier. The whole system then works as a countershaft unit with power coming in on the smaller sun gear and through the "countershaft" (pinions) to the large sun gear. Thus the output shaft turns in the same direction but at a slower speed than the input.

For **high speed,** the direct drive clutch engages and locks the sun gear shaft to the planet pinion carrier. This locks the rotation of the input and output shafts together for a direct drive.

COMPOUND PLANETARY TRANSMISSIONS

Simple planetaries can be combined or connected to obtain extra speed ratios.

To combine two simple planetaries, the output of the first system is connected to the input of the second. Many ratios can then be obtained, as we'll see now.

Compound Planetary Transmission (Twelve Speeds)

Let's build up a compound planetary transmission with eight speeds forward and four speeds in reverse.

NOTE: The brakes will be called B-1, B-2, etc., while the clutches will be named C-1, C-2, etc.

The first two speeds are shown in Fig. 9.

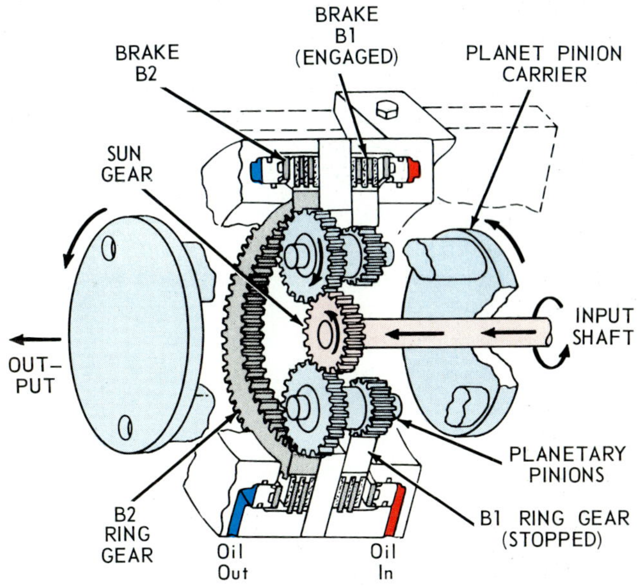

X2160

Fig. 9—Compound Planetary Transmission—Speeds 1 and 2

The planetary consists of an input shaft fastened to a sun gear (shown in pink), compound planetary pinions and carrier (shown in light blue), and two ring gears (shown in gray) each equipped with hydraulic brakes to stop their rotation.

When brake B-1 is applied and the sun gear is turning, the planet pinion carrier is forced to rotate by the smaller of the two planet pinions walking around the B-1 ring gear. This turns the carrier very slowly.

When brake B-2 is engaged, the larger of the planet pinions is forced to walk around the B-2

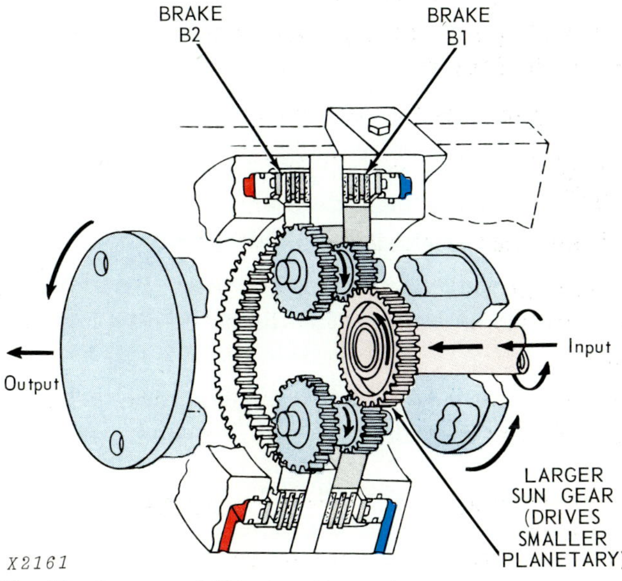

X2161

Fig. 10—Compound Planetary Transmission—Speeds 3 and 4

ring gear. This turns the carrier somewhat faster but still much slower than the input shaft. This gives the two speed ratios both of which are underdrives when compared to the input speed.

For neutral, both B-1 and B-2 brakes are released.

In Fig. 10 we have added a second, larger sun gear on a shaft concentric with the first sun gear shaft. By bringing power into the planetary on a sun gear, two extra speed ratios are obtained by engaging first B-1 and then B-2. These two ratios are slightly faster than the ratios achieved with the small sun gear, but are still slower than the input speed. So far we have four underdrive ratios available.

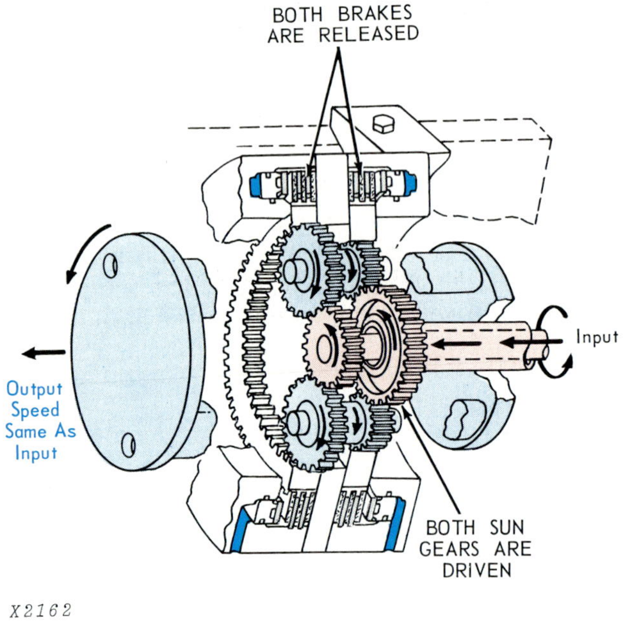

Fig. 11—Compound Planetary Transmission—Speed 5 (Direct Drive)

The fifth speed ratio is a direct drive (Fig. 11). To do this, we drive *both* sun gears at input shaft speeds. (Both brakes must be released.)

Since the sun gears do not have the same number of gear teeth and are in mesh with the planet pinions, they are locked-up and the carrier speed is the same as the input shaft speed.

To direct the flow of power into the two sun gears, two clutches (C-1 and C-2) are installed at the rear of the engine (Fig. 12).

The clutches turn at engine speed whenever the engine is running, ready to transmit power to either one or both sun gears in the planetary. This power is transmitted to the sun gears through two shafts, one inside the other.

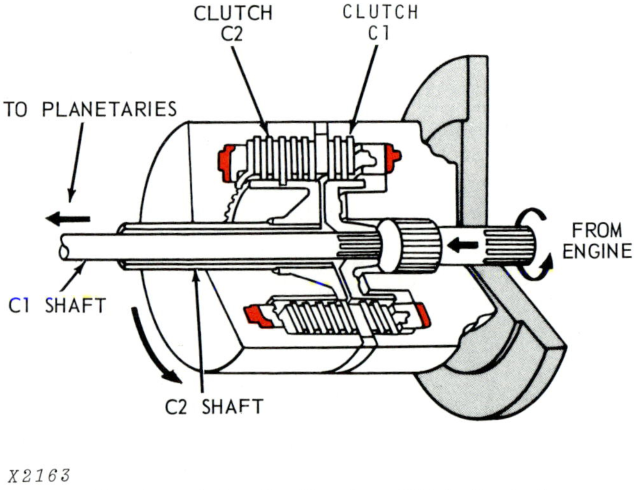

Fig. 12—Hydraulic Clutches Which Control Power To Planetaries

To complete the transmission (Fig. 13), another compound planetary unit (shown in pink) is fastened to the first unit. The planet pinion carriers are fastened together so that the output of the first planetary is the input of the second planetary. This forces the planet pinions to rotate in an orbit around the transmission centerline whenever power is being transmitted through the first planetary. The second planetary is capable of a reverse drive, a direct drive, and an overdrive.

When the B-3 or *reverse* brake is engaged (Fig. 13), the B-3 ring gear stops rotating. As the planet pinions are forced around by the carrier, the large single planet pinion gear is forced to walk around the ring gear and turns in the opposite direction of the carrier. This gear is in mesh with the smaller double planet pinions, and turns them in the same direction as the carrier. This turns the front or largest sun gear to rotate in the opposite direction as the carrier. Since the sun gear is splined to the output shaft, the output shaft turns in the same direction, giving **reverse drive.**

When the C-3 or direct drive clutch is engaged (Fig. 14), the two sun gears in the planetary are forced to rotate together. The hub of C-3 and the first or larger sun gear are both fastened to the output shaft. The C-3 drum is fastened to the second or smaller sun gear. When the two sun gears are locked together by engagement of C-3, the planet pinions cannot turn and the output shaft turns at the same speed as the carrier or input. This is **direct drive.**

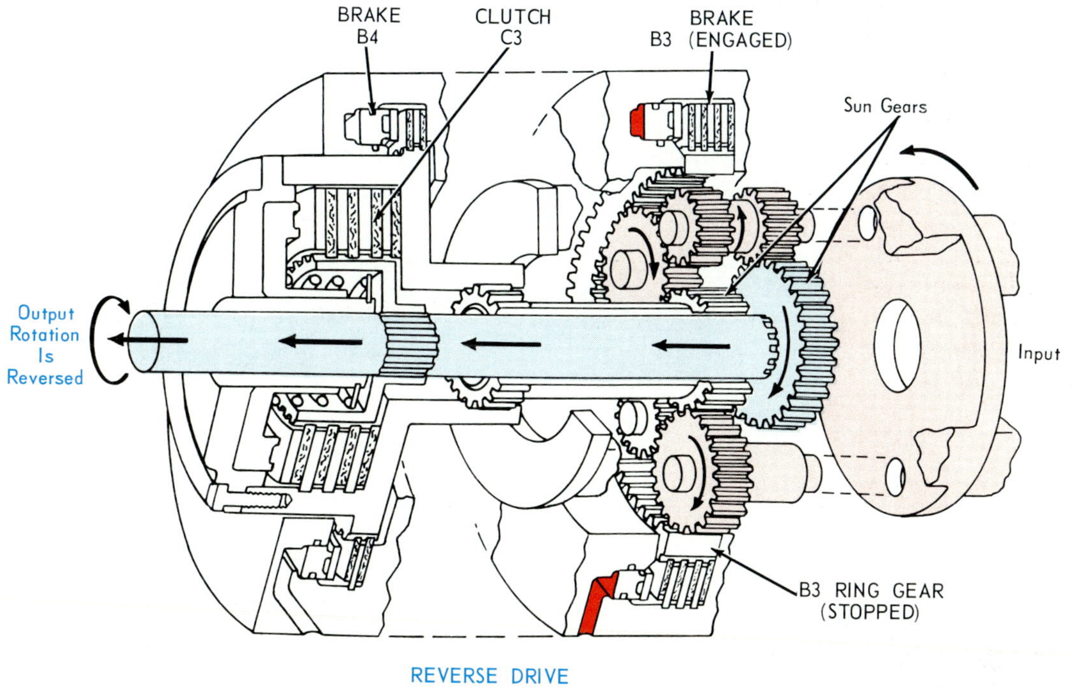

X2164

Fig. 13—Compound Planetary Transmission With Second Planetary Added—Reverse Drive Shown

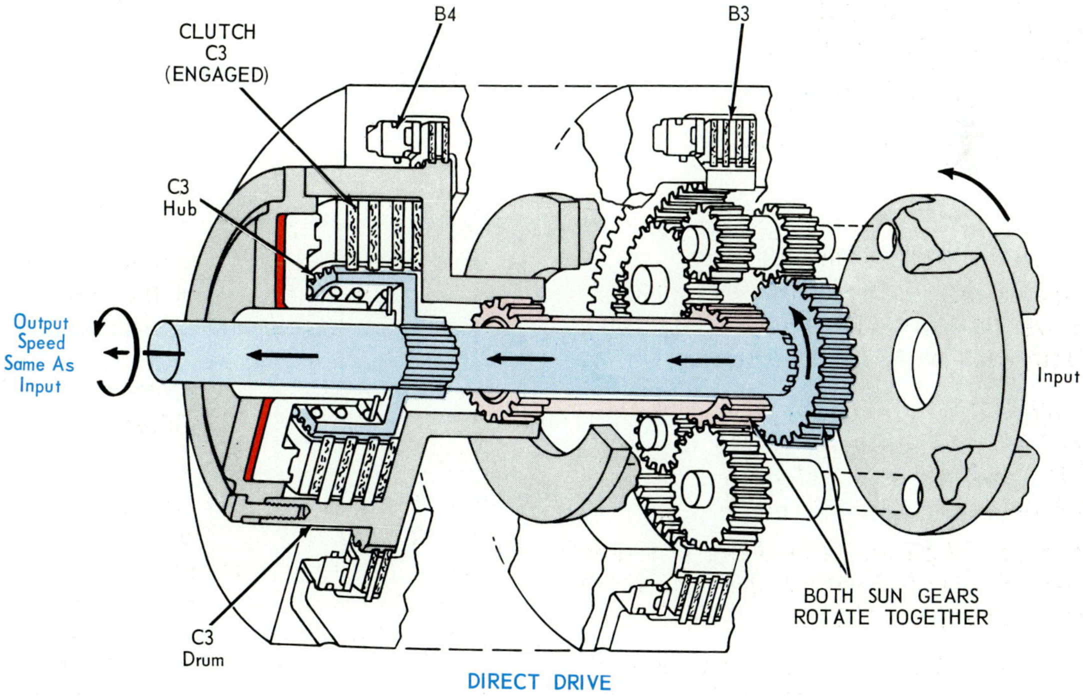

X2165

Fig. 14—Second Planetary Transmitting Direct Drive

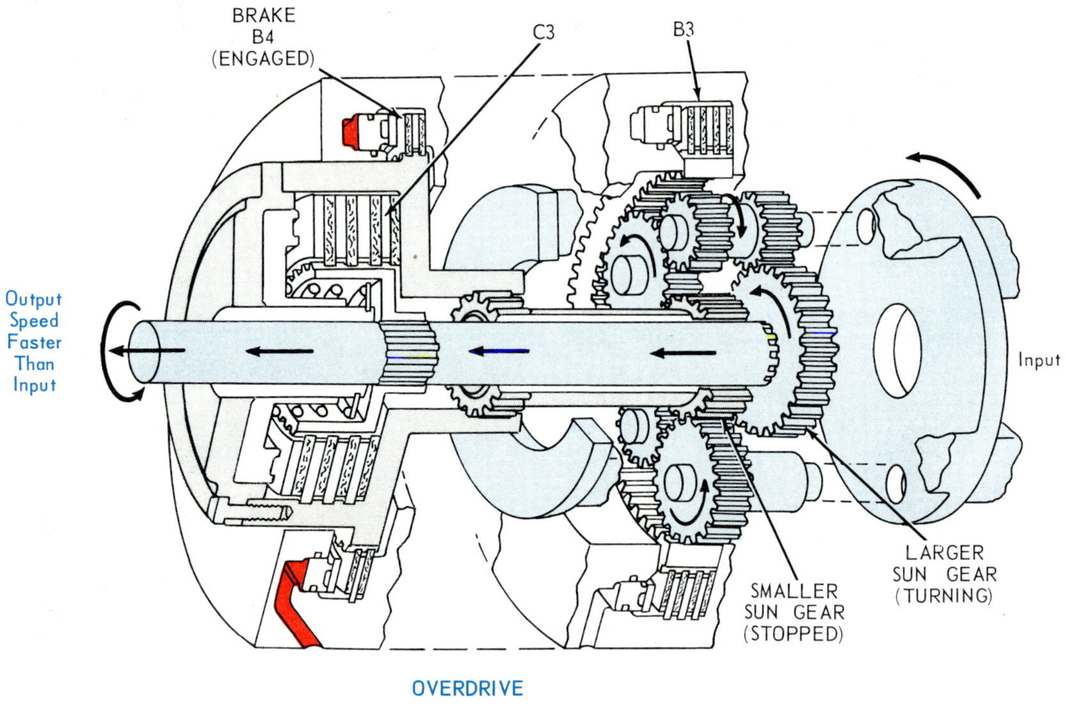

X2166

Fig. 15—Second Planetary Transmitting Overdrive

When the B-4 or overdrive brake is engaged (Fig. 15), the second or smaller sun gear stops rotating, because it is tied to the transmission case through the B4 brake. As the carrier forces the planet pinions around in an orbit, the large single planet pinion is forced to walk around the stationary sun gear. This rotates the large planet pinion the same direction as the carrier, causing the smaller double planet to rotate in the opposite direction. This turns the large sun gear and the output shaft in the same direction as the carrier but at a faster rate of speed. Since the output shaft is turning faster than the input shaft, this ratio is known as **overdrive.**

Since the first planetary is capable of five different ratios and the second planetary is capable of two ratios forward, theoretically the transmission would have 10 speeds forward and five speeds in reverse. However, because of overlap in the ratios, this transmission has 8 forward speeds and 4 reverse speeds, all evenly spaced.

The clutch and brake units that are engaged in each of these ratios is shown in the table (Fig. 16).

An actual compound planetary transmission like the one we have just described is shown in Fig. 17. This unit drives a modern tractor.

1st Fwd.	C1		B1				C3
2nd Fwd.	C1			B2			C3
3rd Fwd.		C2	B1				C3
4th Fwd.		C2		B2			C3
5th Fwd.		C2	B1			B4	
6th Fwd.		C2		B2		B4	
7th Fwd.	C1	C2					C3
8th Fwd.	C1	C2				B4	
1st Rev.	C1		B1		B3		
2nd Rev.	C1			B2	B3		
3rd Rev.		C2	B1		B3		
4th Rev.		C2		B2	B3		

X2167

Fig. 16—Engagement Of Clutches (C) and Brakes (B) In Each Speed Of Compound Planetary Transmission

Summary: Compound Planetary Transmissions

By combining simple planetaries and selecting different parts of each unit as input, output, and

Fig. 17—Compound Planetary Transmission In A Tractor

stationary, a variety of speed ratios can be selected. These range from a large reduction in speed (underdrive) to a large increase in speed (overdrive), plus several reverse drives.

The compound planetary is a very compact transmission for the number of speeds it provides. However, more horsepower is required to run it than a gear-type transmission because of hydraulic drag.

HYDRAULIC SHIFT CONTROLS

Oil does two jobs in a hydraulic assist transmission:

- **Lubricates and cools the clutches and brakes**
- **Engages the clutches and brakes**

Figure 18 on the next page shows an oil circuit for a hydraulic assist transmission. This circuit is an open-center type (see the FOS "Hydraulics" manual).

The transmission case is often used as a reservoir for the oil. The transmission oil pump draws oil through the reservoir screen and into the pump. From the pump, the oil is filtered to remove all contaminates. The filter is protected from rupture by the relief valve. Since an open-center pump is used, a pressure regulator valve is installed in the pump discharge line to regulate and control the pressure.

From this point the oil flows in two directions—the *lubricating oil route* and the *control oil route.*

LUBRICATING OIL ROUTE

The first route is through the pressure regulating valve to the oil cooler to remove the heat and then into the lubricating circuit of the transmission (see pink oil in Fig. 18). This oil both lubricates and cools the transmission. It flushes the clutches and brakes to cool them of the heat built up when the clutch is slipped. (Slipping occurs momentarily every time the unit is shifted, while the clutch is picking up its load.)

CONTROL OIL ROUTE

The second path taken by oil is the clutch control circuit (shown in red in Fig. 18). From the pressure regulating valve this oil flows through an adjustable orifice and fills the spring-loaded accumulator with oil at system pressure. From the outlet of the accumulator, the oil goes to the valves which direct flow to the different clutches. When a valve is shifted to engage a clutch and brake, the accumulator discharges its oil and takes up the volume behind the clutch and brake pistons necessary to remove the clearance in the plates.

Since the accumulator is discharged, the pressure on the clutch and brake pistons at the start of lockup is relatively low and builds up slowly during lockup as the accumulator refills. When the accumulator piston bottoms, the pressure in the clutch and brake jumps to system pressure and completes the clutch lockup or engagement rapidly. If the adjustable orifice is opened wider, the shift is made faster. If the orifice is closed down, the shift is made slower.

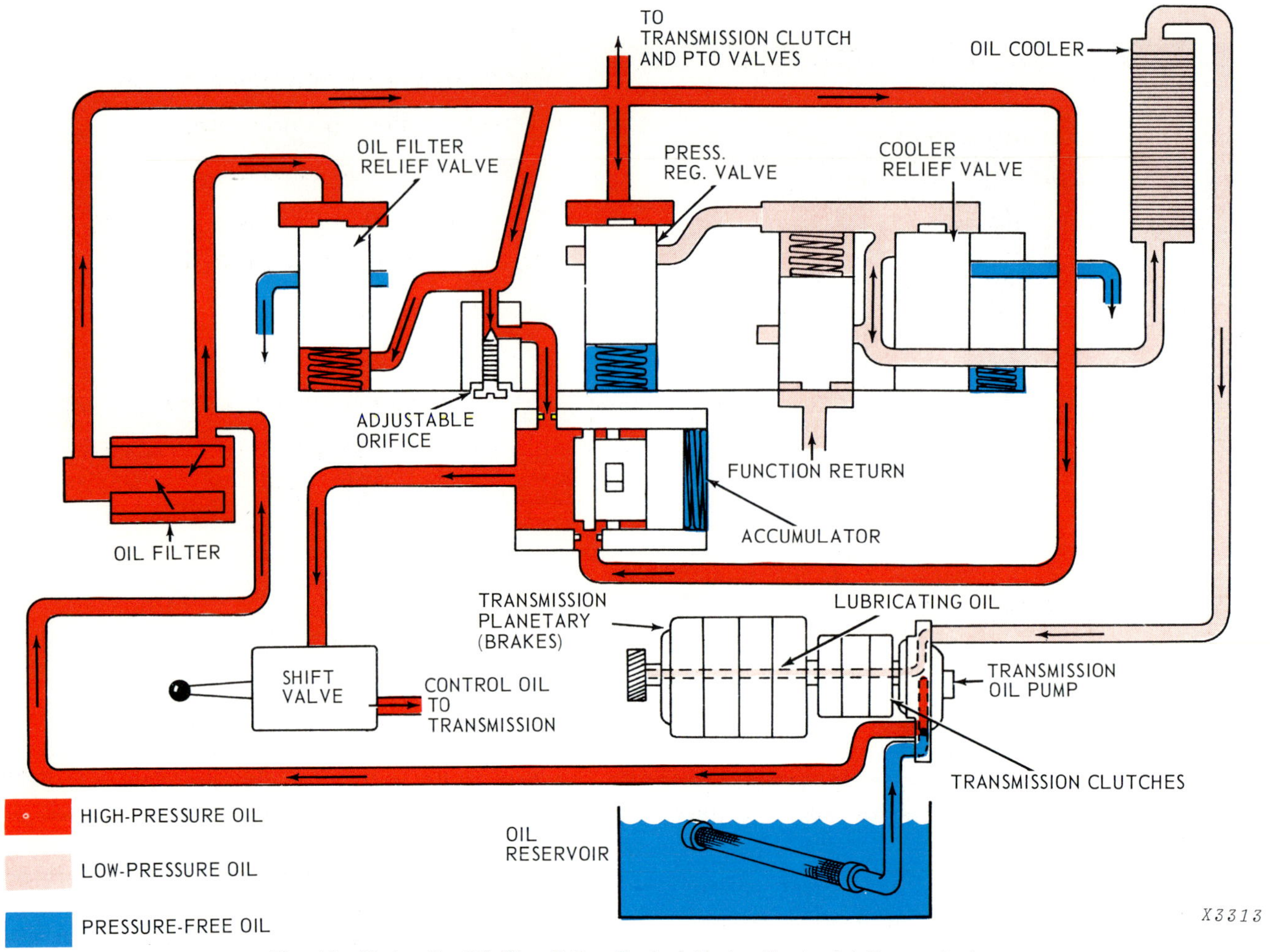

Fig. 18—Hydraulic Oil Circuit For Typical Hydraulic Assist Transmission

This system modulates the shift and prevents the violent lurch that occurs if the clutches are engaged too rapidly.

A *shift valve* is used in the hydraulic assist transmission between the accumulator and the transmission to control the flow of pressure oil to the different clutches and brakes as the operator shifts from one gear to another. The most common shift valves are either the rotary or spool type valve. The shift valve must have as many different positions as there are speed ranges in the transmission. As transmissions become more complex with more speed ranges, the shift valve becomes more complicated. It is often convenient to use two or more valves working together to obtain the necessary ports to control a transmission with many speeds.

NOTE: Operation of basic hydraulic components is covered in the FOS "Hydraulics" manual.

OTHER CONTROL FEATURES

Three extra features are common in hydraulic assist transmissions:

1. An **inching pedal** which disengages enough of the elements in the transmission to obtain a neutral in any gear ratio. This provides a pedal for emergency stops and for maneuvering the machine in close quarters or when attaching equipment.

2. A dry-type **disconnect clutch** between the engine and the transmission for cold weather starting. Cold, stiff oil between the plates causes all the clutches and brakes in the transmission to drag, causing a lockup and drag on the engine starting circuit.

3. A **tow disconnect** or *neutral* between the transmission and the final drive members is necessary if the machine is to be towed without the engine running. Since the transmission pump works only when the engine is running, no lubricating oil is available while towing. As a result, some of the shafts and clutches rotating without lubrication would be damaged—unless they can be disconnected. The tow disconnect is normally provided by a splined sliding collar which disconnects the reduction gear from the reduction gear shaft.

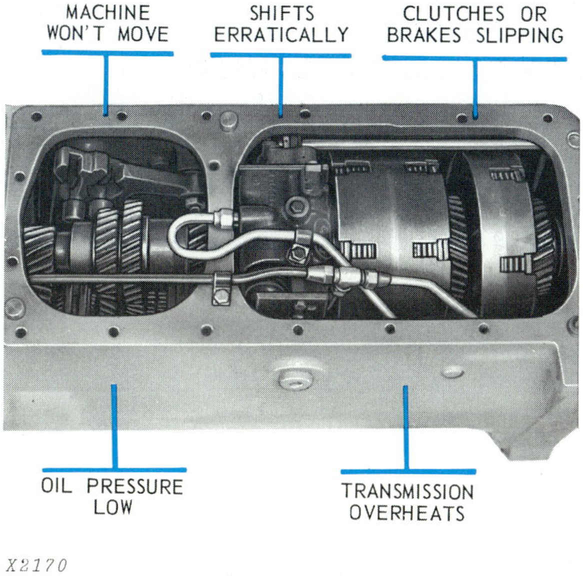

Fig. 19—Troubleshooting Problems With Hydraulic Assist Transmissions

TROUBLESHOOTING

These troubleshooting charts are given as a general guide to the common failures of hydraulic assist transmissions.

Supplement these charts with specific trouble shooting from the machine Technical Manual.

MACHINE WON'T MOVE

1. Tow disconnect disengaged.
2. Cold weather starting clutch disengaged.
3. Failed clutch.
4. Park lock engaged.
5. No oil pressure.
6. Control linkage stuck or disconnected.
7. Blown packing or gaskets.
8. Oil filter plugged.

SHIFTS ERRATICALLY

1. Wrong setting of adjustable orifice.
2. Incorrect system oil pressure.
3. Failed or stuck accumulator.
4. Shift control disconnected or binding.

CLUTCHES OR BRAKES SLIPPING

1. Low oil pressure.
2. Defective gasket or seal.
3. Burned or warped clutch facings.
4. Excessive oil leakage at shift valves.

LOW SYSTEM PRESSURE

1. Plugged oil filter.
2. Low oil level.
3. Blown O-ring or gasket.
4. Plugged reservoir oil screen.
5. Air leak in pump suction line.
6. Wrong adjustment of pressure regulating valve.
7. Inching pedal valve leaking.
8. Stuck filter bypass valve.
9. Stuck cooler relief valve.
10. Failed or worn pump.
11. Pump not turning.

TRANSMISSION OVERHEATING

1. Reservoir oil level too low or too high.
2. Plugged oil filter.
3. Plugged core in oil cooler.
4. Not enough lubricating oil flow.
5. Faulty temperature gauge.
6. Clutch or brake dragging or slipping.
7. Cooler relief valve stuck open.

Litho in U.S.A.

TEST YOURSELF

QUESTIONS

1. (Fill in the blanks.) "In hydraulic assist transmissions, _______ _______ **control** *the power flow, while the _______ _______* **transmits** *the power flow."*

2. What are the two basic types of gear sets used in hydraulic assist transmissions?

3. What does "underdrive" and "overdrive" mean?

4. Name the three main components of a planetary gear set.

5. True or false? "When any two parts of a planetary are locked together, all three parts rotate as a unit."

6. What is the gear ratio of input to output when a planetary has two parts locked together?

7. What two jobs does oil do in a hydraulic assist transmission?

(Answers on page 9 in back of text.)

 Litho in U.S.A.

HYDROSTATIC DRIVES / CHAPTER 5

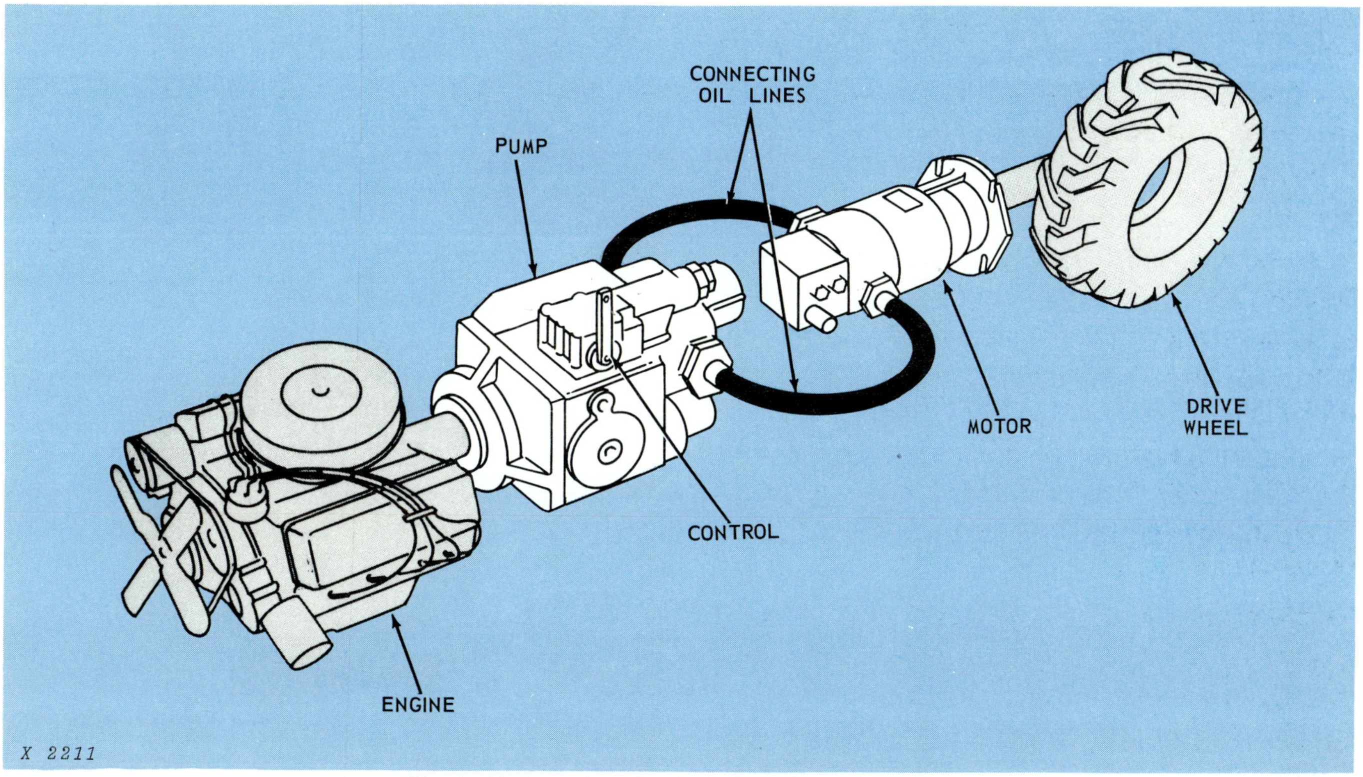

Fig. 1—Hydrostatic Drive In A Complete Power Train

INTRODUCTION

The hydrostatic drive is a fluid drive which uses fluid under pressure to transmit engine power to the drive wheels of the machine.

Mechanical power from the engine is converted to hydraulic power by a pump-motor team. This power is then converted back to mechanical power for the drive wheels.

The hydrostatic drive can function as both a clutch and transmission. The final gear train can then be simplified, with the hydrostatic unit supplying infinite speed and torque ranges as well as reverse speeds.

HOW IT WORKS

There are two basic types of hydraulic transmissions:

- **Hydrodynamic**
- **Hydrostatic**

HYDRODYNAMIC TRANSMISSIONS (Fig. 2) use fluids at *high speeds* but relatively *low pressures.* Basically, a hydraulic pump (hose) forces fluid against a turbine wheel, rotating it as shown.

This type of transmission is used in **torque converters** (Chapter 6).

HYDROSTATIC TRANSMISSIONS (Fig. 2) use fluids at *high pressures* but relatively *low speeds.* Basically, energy is transferred by the fluid itself in a closed circuit between the pump and motor as shown. While the fluid does move through the lines (shown in red), it is still considered as being at rest or under *static* pressure. The rise in pressure of the fluid—which will not compress—is what transfers the energy.

This type of transmission is used in the **hydrostatic drives** we'll cover in this chapter.

 Litho in U.S.A.

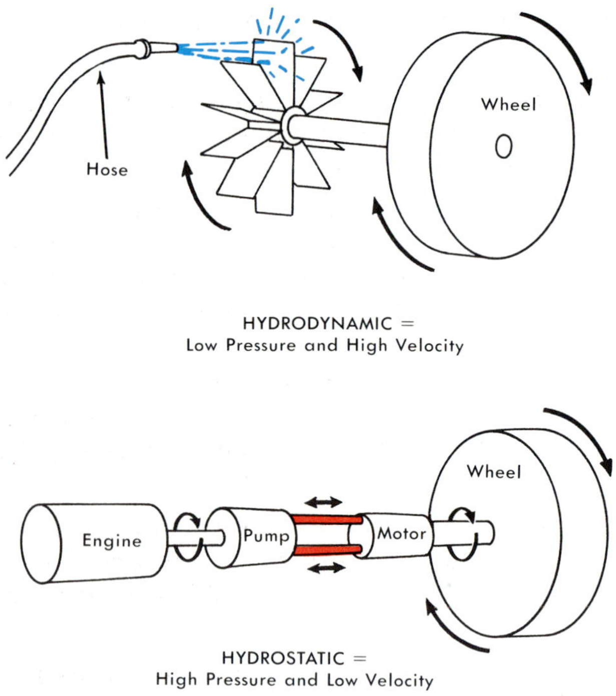

X 2212

Fig. 2—Basic Types Of Hydraulic Transmissions

THEORY OF HYDROSTATIC DRIVES

Two principles of hydraulics are required to understand a hydrostatic drive:

- **Liquids have no shape of their own**
- **Liquids will not compress**

X 1071

Fig. 3—Liquids Have No Shape Of Their Own

LIQUIDS HAVE NO SHAPE OF THEIR OWN. They acquire the shape of any container (Fig. 3). Because of this, oil in a hydraulic system will flow in any direction and into a passage of any size or shape.

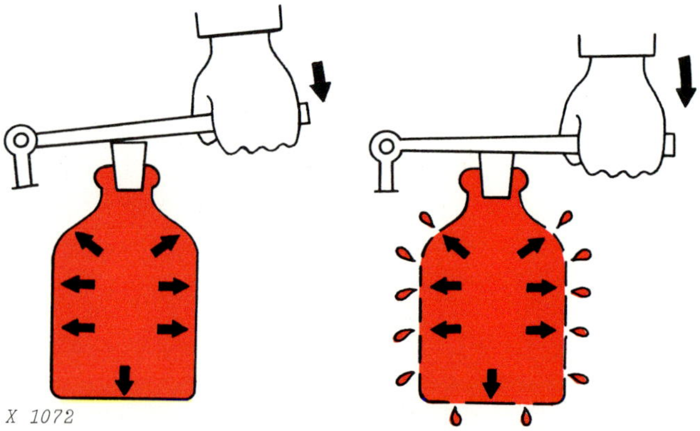

X 1072

Fig. 4—Liquids Will Not Compress

LIQUIDS WILL NOT COMPRESS. Fill a plastic jug with liquid (Fig. 4). Then tightly cork it. Now let's say you pushed down on the cork and tried to compress the liquid. It wouldn't work—the jug would expand and eventually burst.

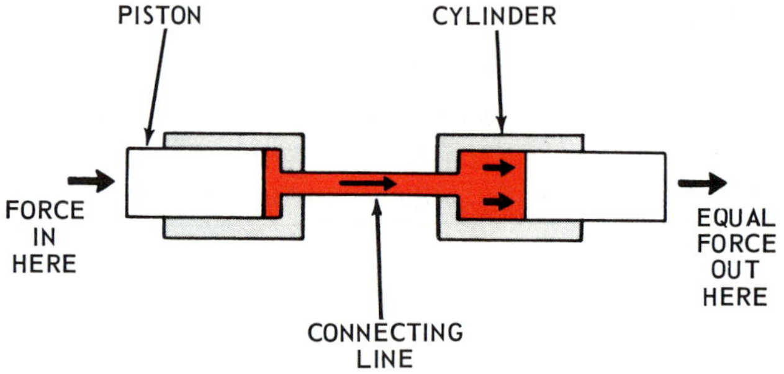

X 2213

Fig. 5—Two Cylinders Connected

Fig. 5 shows how these principles work in a hydrostatic drive.

Two cylinders, each containing a piston, are connected by a line. The cylinders and the line are filled with oil.

When a force is applied to the left piston as shown, that piston moves against the oil. The oil will not compress, so it acts as a solid and forces out the right piston.

In a hydrostatic drive, several pistons are used to transmit power—one group in the **pump** sending power to another group in the **motor.**

The pistons are in a cylinder block and revolve around a shaft. The pistons also move in and out of the block parallel to the shaft as we'll see later.

 Litho in U.S.A.

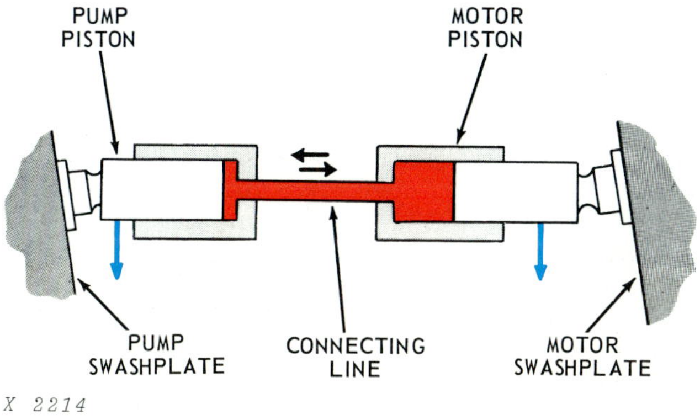

Fig. 6—Two Connected Cylinders With Swashplates

Fig. 6 shows one piston for the pump and one for the motor. To provide a pumping action for the pistons, a plate called a **swashplate** is located in both the pump and motor. The pistons ride against the swashplates.

The angle of the swashplates (Fig. 6) can be varied so that the volume and pressure of oil pumped by the pistons can be changed or the direction of oil flow reversed.

A pump or motor with a movable swashplate is called a **variable-displacement** unit. A pump or motor with a fixed swashplate is called a **fixed displacement** unit.

Fig. 7 shows a variable displacement pump driving a fixed displacement motor.

As the pump pistons rotate, they move across the sloping face of the swashplate, sliding in and out of their cylinder bores to pump oil out. The more the pump swashplate is tilted, the more oil it pumps with each piston stroke and the faster it drives the motor.

Pressure results whenever the flow of a fluid is resisted. The resistance may come from acceleration of the machine or normal load.

The motor swashplate is at a fixed angle so that the stroke of its pistons are always the same. Thus its speed of rotation cannot be changed except as it is driven faster or slower by the pump oil.

The point to remember now is that a given volume of oil forced out of the pump will cause the motor to turn at a given speed. More oil will speed up the motor; less oil will slow it down.

The pump is driven by the machine's engine and so is linked to the speed set by the operator. It pumps a constant stream of high-pressure oil to the motor.

Since the motor is linked to the drive wheels of the machine, it gives the machine its travel speed.

Only three factors control the operation of a hydrostatic drive:

- **Rate of oil flow—gives the speed**
- **Direction of oil flow—gives the direction**
- **Pressure of oil—gives the power**

Control of these three factors is infinite, giving endless selections of speed and torque in a hydrostatic drive.

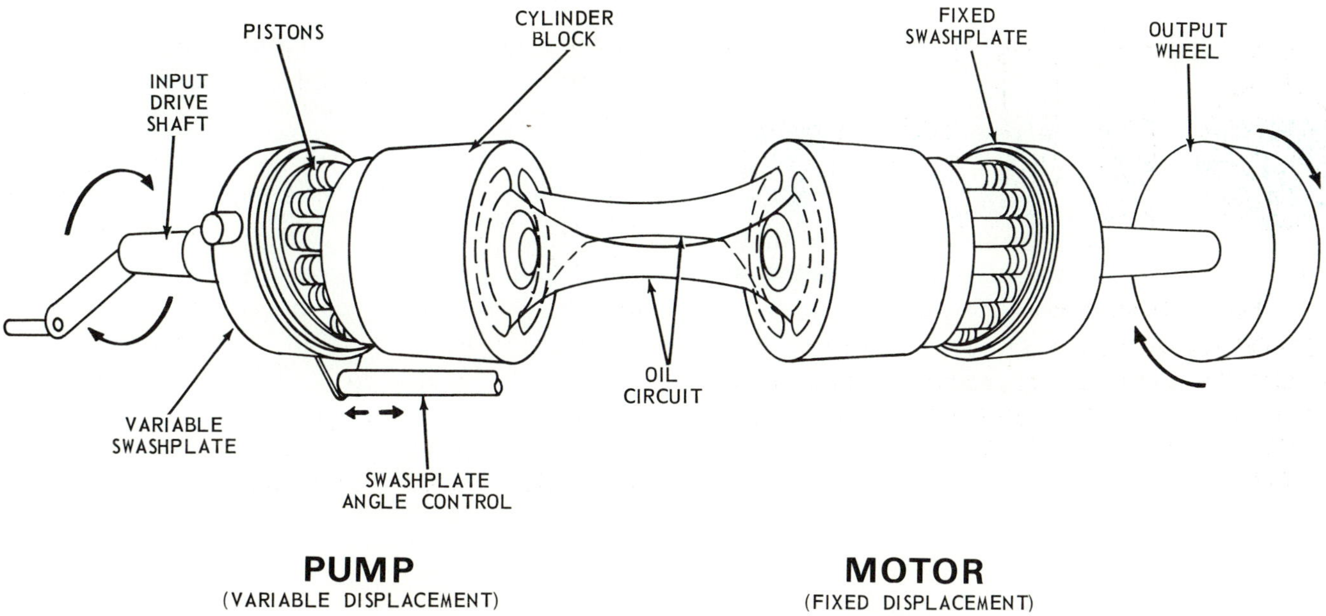

Fig. 7—Variable Displacement Pump Driving Fixed Displacement Motor

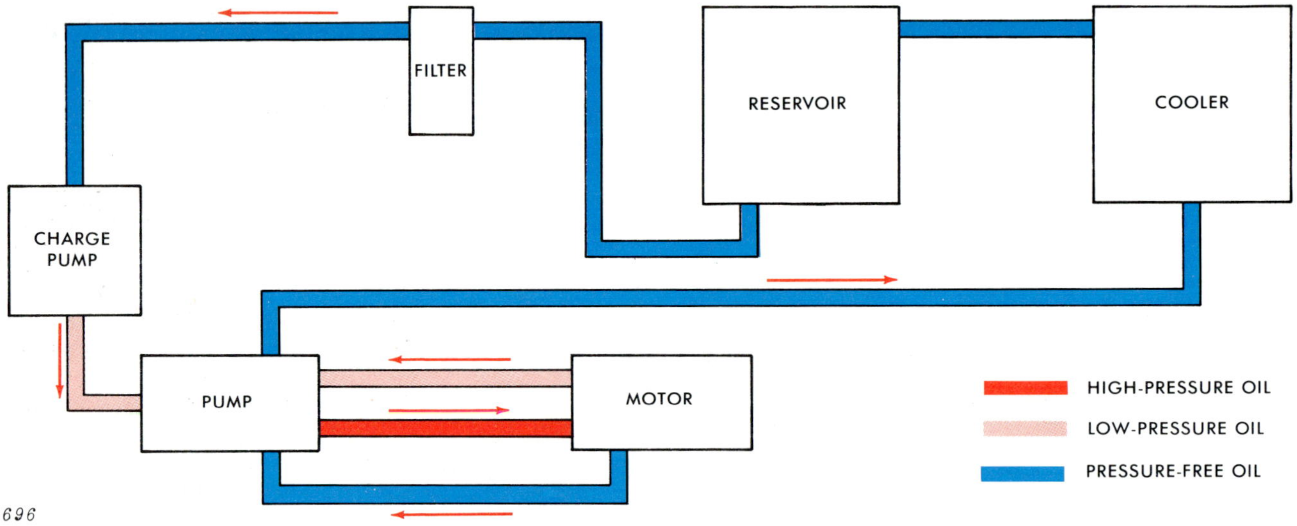

Fig. 8—Complete System For A Hydrostatic Drive (Closed Hydraulic Loop)

The pump-motor team is the heart of the hydrostatic drive, although the complete hydraulic system (Fig. 8) also includes a **reservoir** to supply the oil, a **filter** to remove dirt, and a **cooler** to remove excess heat from the oil.

Basically however, the pump and motor are joined in a closed hydraulic loop; the return line from the motor is joined directly to the intake of the pump, rather than to the reservoir. The charge pump simply supplies the oil, drawing it from the reservoir.

NOTE: For basic theory of hydraulic systems, see the FOS "Hydraulics" manual.

TYPES OF HYDROSTATIC DRIVES

Displacement is the quantity of fluid which a pump can move (or a motor can use) during each revolution. It is directly related to the horsepower output of the drive. (Horsepower is a combination of torque X speed.)

As you have already learned, pumps and motors can have a **fixed** displacement or **variable** displacement.

Four pump-motor combinations are possible:

1) **Fixed** *displacement pump driving a* **fixed** *displacement motor.*

2) **Variable** *displacement pump driving a* **fixed** *displacement motor.*

3) **Fixed** *displacement pump driving a* **variable** *displacement motor.*

4) **Variable** *displacement pump driving a* **variable** *displacement motor.*

The point to remember is that for each combination the output power must equal the input power minus negligible power losses. Let's look at each combination (Fig. 9).

No. 1 circuit is like a gear drive; it transmits power without altering the speed or horsepower between the engine and the load. A constant input speed and torque gives a constant output horsepower. If either the speed or torque are increased holding the other constant the output horsepower will increase.

Since the pump flow is variable in *No. 2 circuit,* output speed is variable and torque output is constant for any given pressure. This circuit gives variable speed and constant torque.

Output speed in *No. 3 circuit* is varied by changing the motor displacement. For constant power input if the motor displacement decreases, output speed increases but output torque drops.

No. 4 circuit is the most flexible and expensive of all the circuits but also the most difficult to control. It is capable of operating like any of the above combinations.

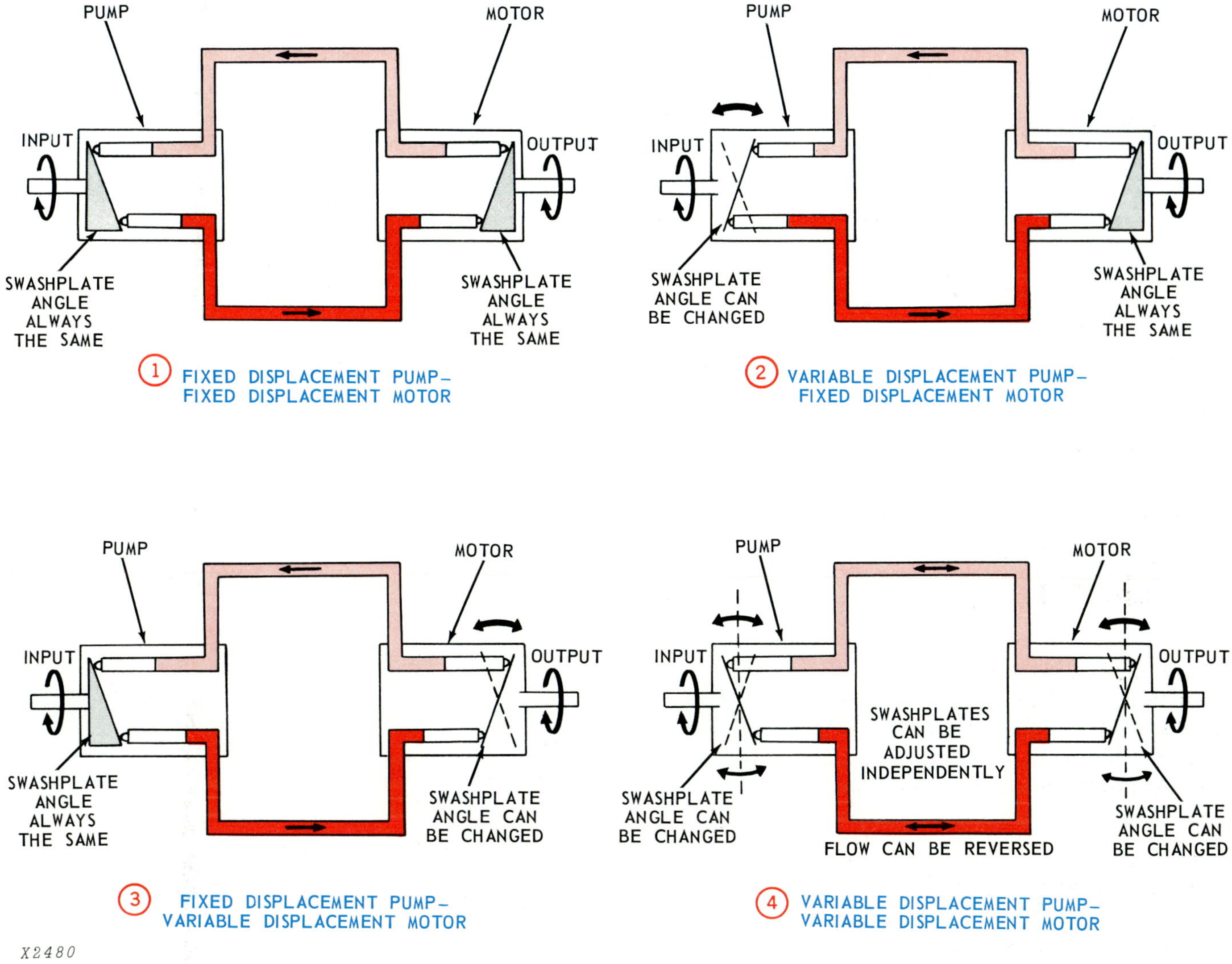

Fig. 9—Pump-Motor Combinations For Hydrostatic Drives

REVERSING IN VARIABLE PUMP-MOTOR CIRCUITS

The direction of output shaft rotation can be reversed in variable circuits by shifting either the pump or motor swashplates over center (Fig. 10).

In **neutral,** the swashplate is vertical and no oil is pumped.

In **forward,** the swashplate is tilted and oil is pumped as shown at the top.

In **reverse,** the swashplate is tilted the opposite way and the unit pumps oil in the opposite direction.

How this affects the complete drive will be seen under "Operation", which follows.

 Litho in U.S.A.

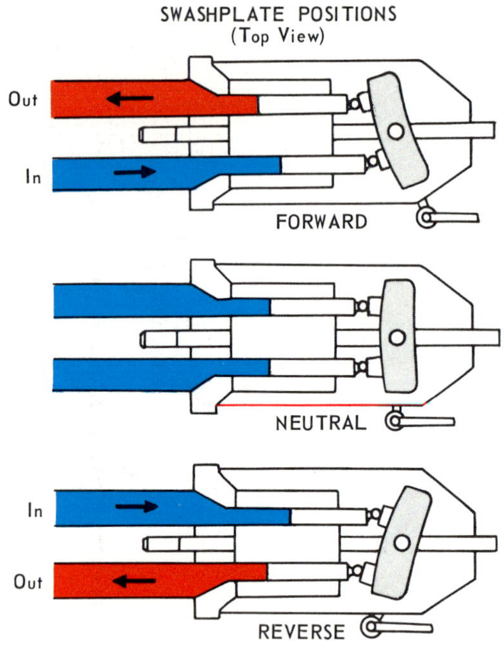

X 2219
Fig. 10—How Reversing Is Done With Variable Displacement Pumps Or Motors

OPERATION OF HYDROSTATIC DRIVES

Let's see how a typical hydrostatic drive works.

Our example is a system with an axial piston *pump* (Fig. 11) and *motor* (Fig. 12). This type of unit is most common in hydrostatic drives today.

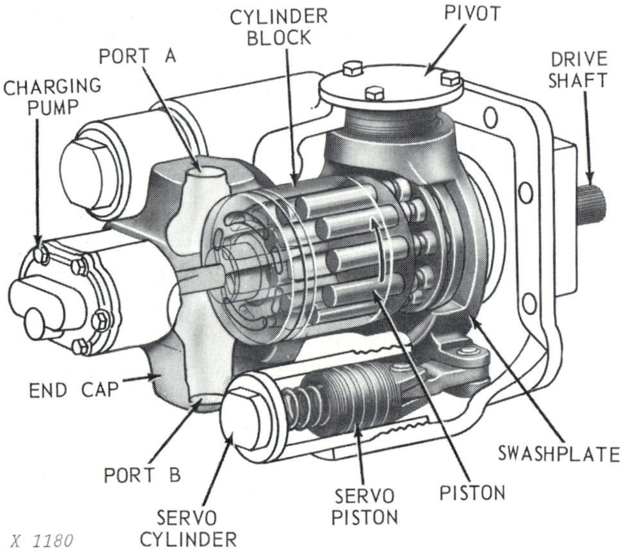

X 1180
Fig. 11—Variable Displacement Pump—Axial Piston Type

The *pump* has a **variable** displacement, while the *motor* has a **fixed** displacement. (These units are shown connected in Fig. 7).

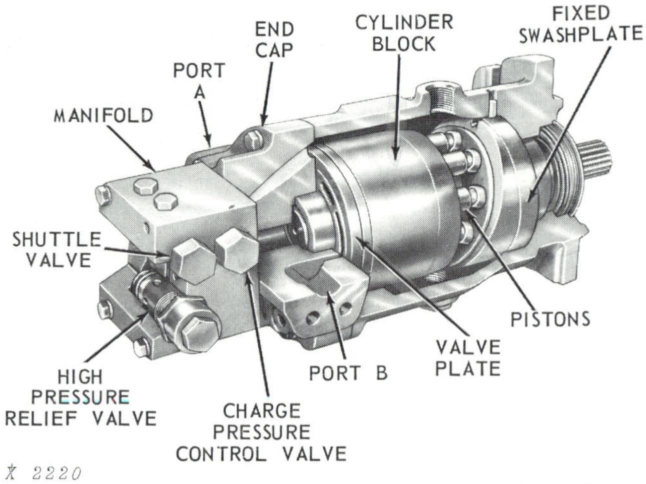

X 2220
Fig. 12—Fixed Displacement Motor—Axial Piston Type

Let's look at the complete system in operation in *neutral, forward,* and *reverse* speeds.

NEUTRAL (Fig. 13)

When the speed range control lever is in neutral, free oil flows from the reservoir through the oil filter to a charge pump. This pump (which is a gear-type to provide a constant supply of oil to the main pump) pumps the oil into the center section where it circulates through the motor housing. Oil then passes through the center section, circulates throughout the pump housing, and then returns through the oil cooler to the reservoir.

Trapped oil (shown in green) is held in the cylinder block of the pump, in the motor, and in the connecting lines between the pump and motor by two check valves (A) and (B) in the center section.

When the control lever is in neutral, the swashplate in the pump is also in neutral and the pistons in the pump are not stroking. Therefore, no oil is being pumped.

The cylinder block in the pump, driven by the machine engine rotates in a counterclockwise direction as viewed from the input end of the pump.

Because oil is not being pumped to the motor, the cylinder block in the motor is stationary and the output shaft does not move.

The hydrostatic belt tightener valve provides for pressure oil to assist a spring to keep proper tension on the belt.

 Litho in U.S.A.

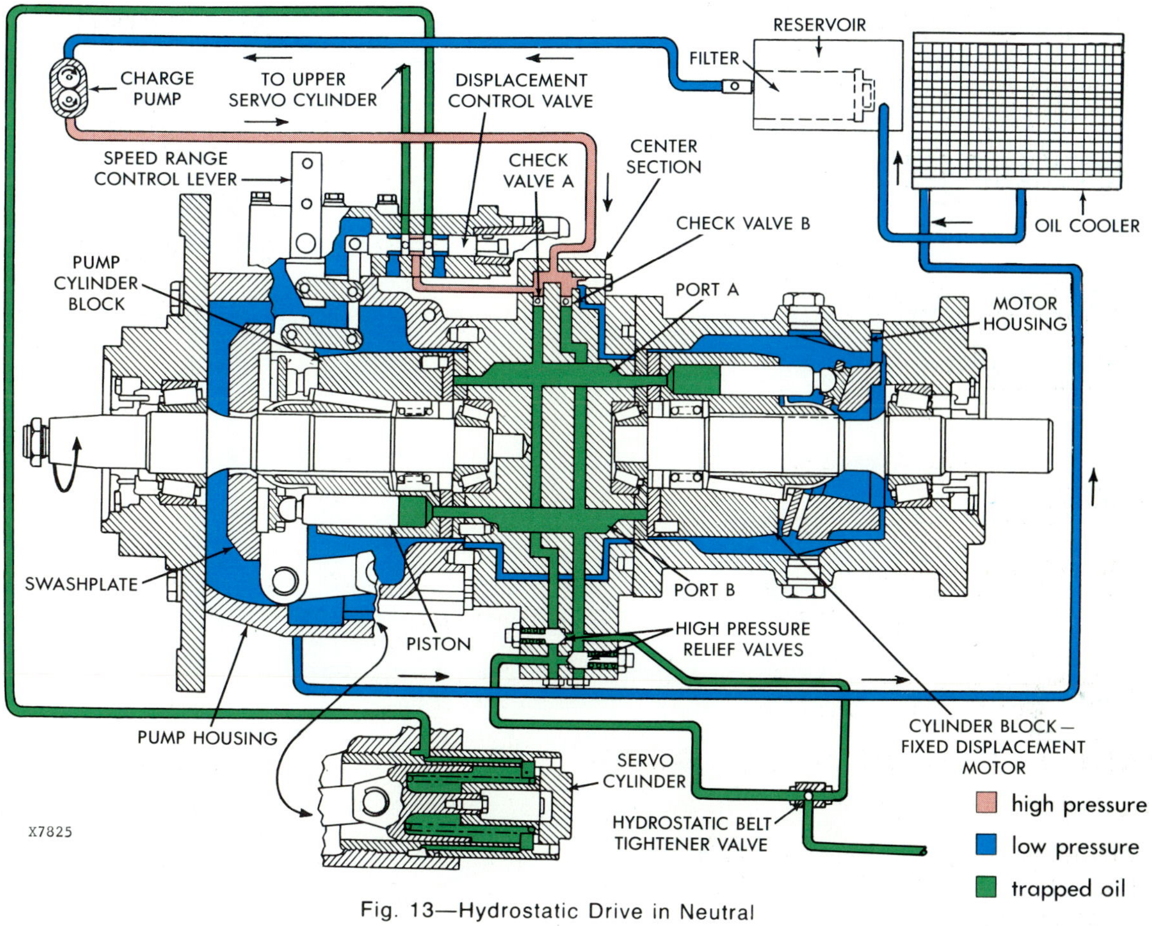

Fig. 13—Hydrostatic Drive in Neutral

FORWARD (Fig. 14)

When the speed range control lever is moved forward, the spool in the displacement control valve moves out of neutral. This allows oil to flow into the lower servo cylinder forcing the swashplate (which is connected to the servo cylinder) to tilt. Oil expelled by the opposing servo cylinder (not shown) returns through the displacement control valve to the pump case.

When the swashplate reaches the tilt set by the control lever, the displacement control spool returns to neutral (as shown in Fig. 14). This traps the oil to both servo cylinders and holds the swashplate tilted as shown. The swashplate will remain in this position until the speed control lever is moved by the operator.

With the pump drive shaft and cylinder block rotating counterclockwise and the top of the swashplate tilted to the left, low pressure oil enters port A through check valve A. This forces the pistons in the pump, that align with port A, to go out against the swashplate, filling the pump cylinder block. The further the swashplate is tilted, the more oil the pump receives. As the driveshaft continues to turn the cylinder block, the pistons are forced back into the block because of the tilted position of the swashplate. As the pistons move back into the block, oil in the cylinder becomes pressurized. Then, when the cylinder is aligned with port B, pressure oil moves through port B into the cylinder in the motor that is aligned with port B.

The distance the pistons in the pump reciprocate in and out of the cylinder block depends on the angle the pump swashplate is tilted. This determines the volume of oil displaced per revolution of the pump. The greater the angle, the greater the volume and the more oil flows from the pump. Thus, as the swashplate angle is changed, the volume of oil displaced from the pump will also change. The greater the angle, the greater the volume, hence a greater flow of oil from the pump.

As pressure oil moves from pump to motor through port B, pistons in cylinders in the motor that align with port B are pushed against the fixed swashplate. The pistons slide down the inclined surface and resulting forces cause the cylinder block to rotate which in turn rotates the output shaft. This drives the differential and transmission causing the machine to move forward.

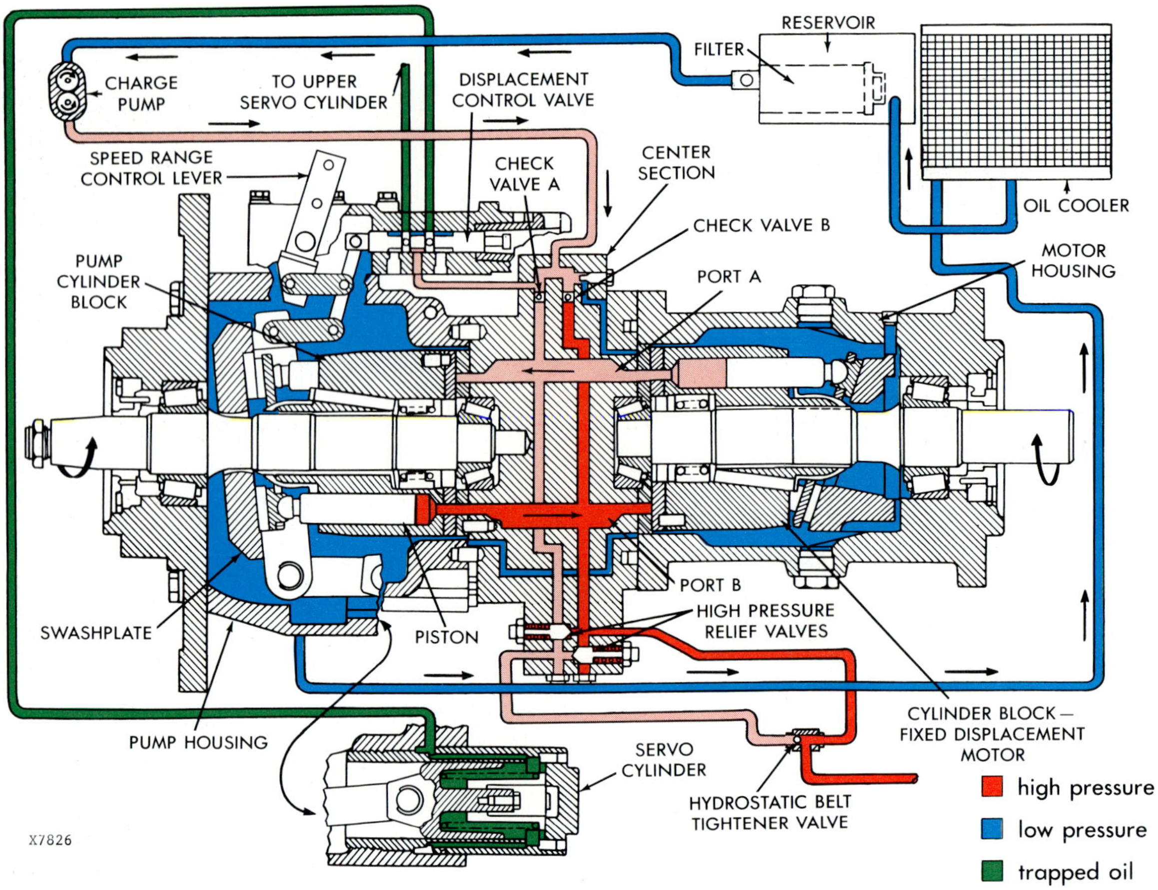

Fig. 14—Hydrostatic Drive in Forward Operation

The cylinder block in the motor rotates in a clockwise direction (viewed from the output shaft end of the motor) when traveling forward. As the cylinder block continues to rotate, oil is forced out of port A, through the center section, to the pump.

The oil returned by the motor to the pump through the center section is recirculated through the pump cylinder block assembly and back to the motor. This is called a "closed circuit" system in that the oil continually circulates between the pump and the motor. The only oil introduced into the closed circuit from the reservoir comes from the charge pump.

If pressure exceeds the rated pressure of the high pressure relief valves, the relief valve opens and oil bypasses the cylinder block assembly in the motor, slowing or stopping the combine. The bypassing oil returns to the pump.

REVERSE (Fig. 15)

When the speed range control lever is moved to reverse, the spool in the displacement control valve moves out of neutral. This allows pressure oil to flow into the upper servo cylinder, tilting the swashplate as shown.

When the swashplate reaches its desired tilt, as set by the control lever, the displacement control spool returns to neutral (as shown in Fig. 15). This traps the oil to both servo cylinders and holds the swashplate tilted as shown. The swashplate remains in this position until the speed control lever is moved again by the operator.

With the pump drive shaft and cylinder block rotating counterclockwise and the top of the swashplate tilted to the right, port B becomes the *inlet* and port A the *outlet* (high pressure port). As the cylinder rotates past port B, check valve B opens and oil is forced by the charge pump into the piston bores, which align with port B. Then as rotation continues, the oil is pressurized and forced out of port A by each of the pistons as they align with port A. This oil flows to the motor.

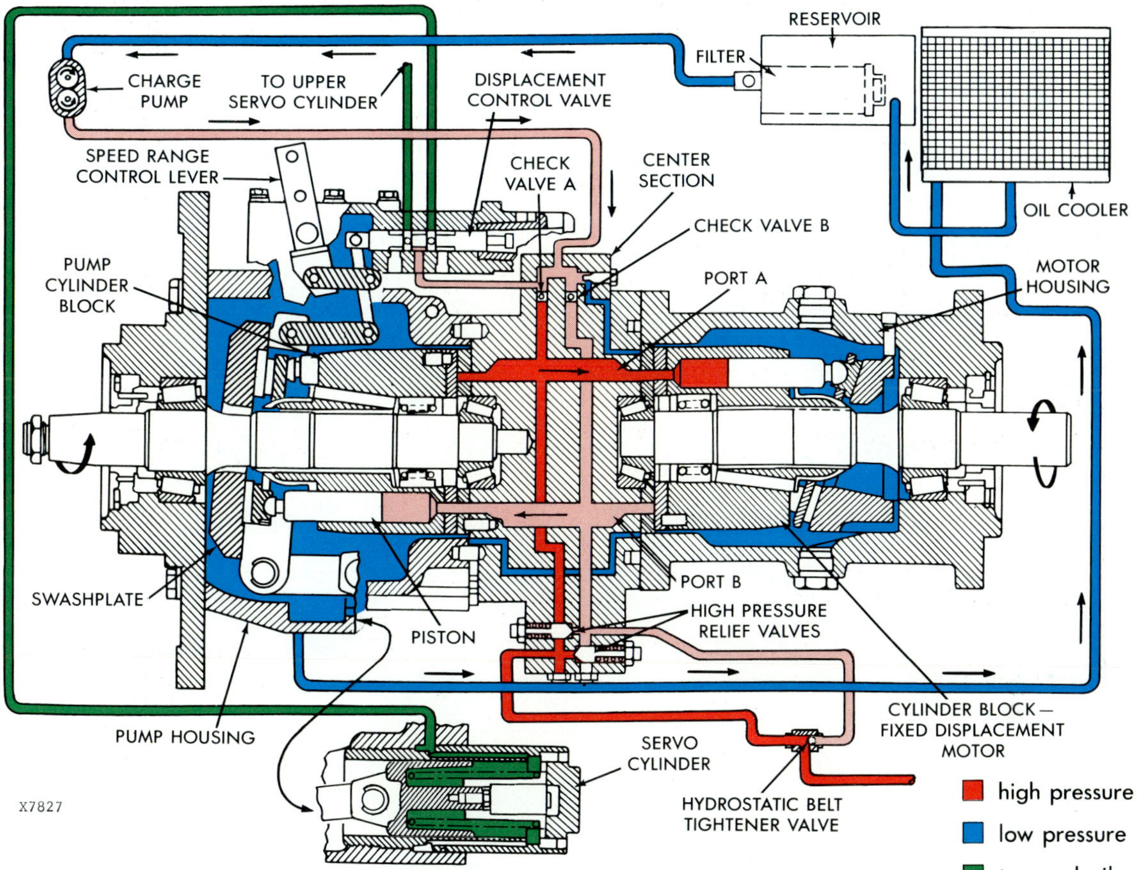

Fig. 15—Hydrostatic Drive in Reverse Operation

As high pressure oil from the pump enters port A of the motor, it pushes the motor pistons against the swashplate. The pistons slide down the inclined surface of the swashplate, rotating the cylinder block which in turn rotates the drive shaft, driving the machine—but in reverse.

As the motor cylinder block continues to rotate, oil is forced out port B at low pressure and returns to the pump.

NOTE: The **pump** *drive shaft and cylinder block always rotate clockwise, but the* **motor** *drive shaft and cylinder block can rotate in either direction, depending on the direction of oil entering from the pump.*

HYDROSTATIC DRIVE AXLES

Integral pump-motor units can be used, one on each drive axle, to propel some machines (Fig. 16).

Each pump-motor is connected at right angles. Both pump and motor have a variable displacement, giving a variable torque and horsepower.

Closed-circuit oil paths between the pump and motor are built into the housing.

A charge pump is built into the unit shown, as well as pressure control and relief valves. The charge pump is a vane type with a fixed displacement.

The swashplates are controlled by linkages which fasten to the yokes above the pump and motor.

The complete system is shown in Fig. 17.

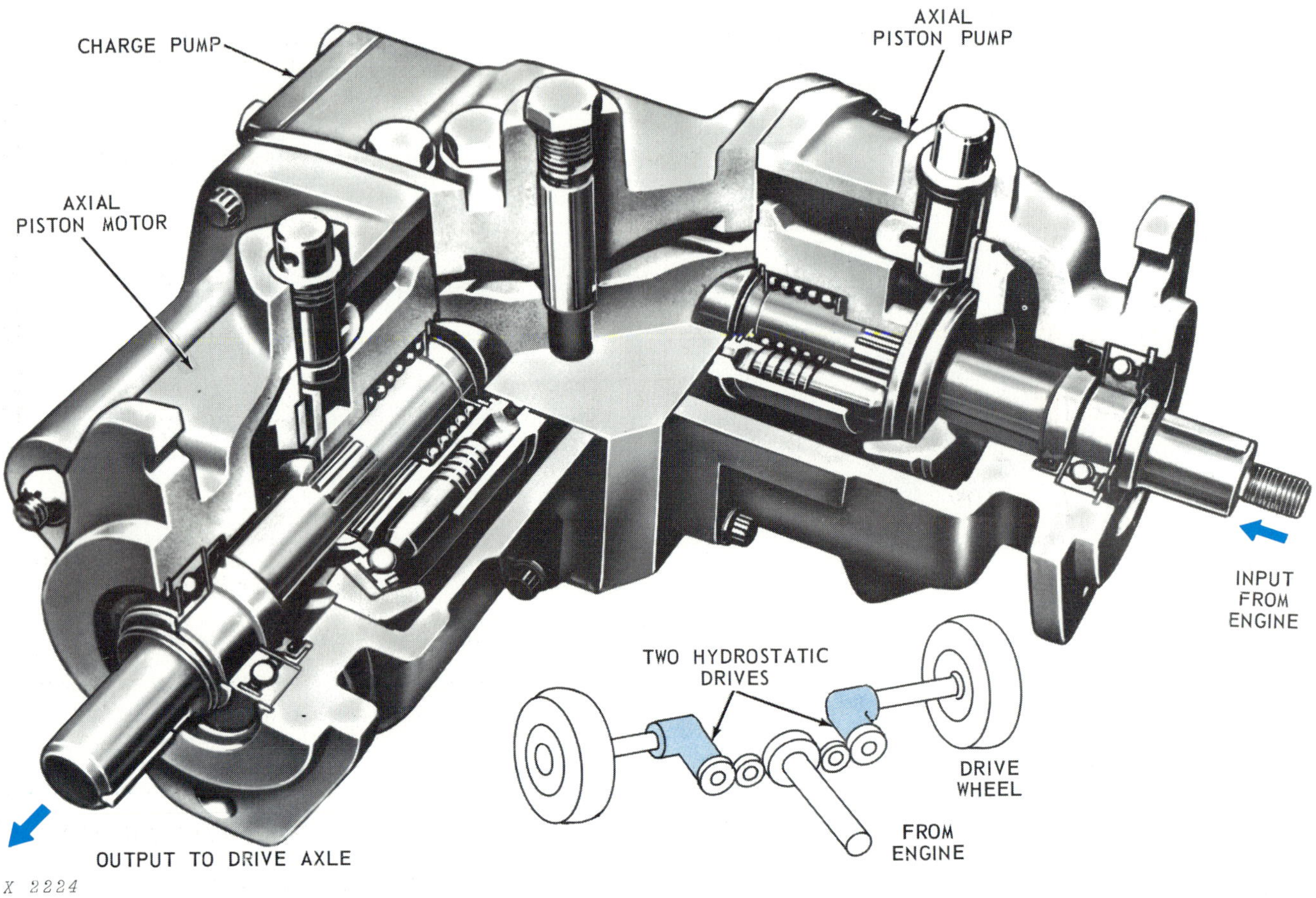

Fig. 16—Integral Pump-Motor For Hydrostatic Drive Axle

OPERATION

Fig. 17 illustrates the oil flow from the **reservoir,** to one **hydrostatic axle,** and the return to reservoir.

Oil flows from the **reservoir** to the **inlet port** of the **charge pump.** The oil passes from the charge pump to the **control valve.** If the control valve is not actuated, oil passes through the control valve into a common line with oil from the other control valve, back to the **filter.** After oil passes through the filter and **cooler,** it enters the **valve block** inlet port of both drives.

The valve block has common oil passages to allow oil to flow from the **piston pump** to the **piston motor** and return to the piston pump. This is a *closed circuit.* When any leakage occurs, the circuit is replenished by cooled oil entering the valve block.

The replenishing oil overcomes the 10 psi (69 kPa) spring setting and discharges into the inlet side (low pressure side) of the piston pump at a pressure in excess of 10 psi (69 kPa). Any excess pressure in the common passage empties over the 25 psi (172 kPa) relief valve into the motor housing and back to the reservoir. The 25 psi (172 kPa) relief valve consists of a sleeve poppet and spring. Pressure acts directly on the poppet against the spring.

To take care of the pressure surges in the outlet passages, relief valve action is provided. The replenishing and relief valve cartridges are pre-set.

The relief valve operates when the outlet pressure on the end of the inner poppet creates a force greater than the pilot spring force (4500 psi) (31 025 kPa) and unseats the inner poppet. This reduces pressure in the spring chamber until the outer poppet moves, discharging excess pressure to the relief cavity. The inner poppet is held open by its pin until the system pressure drops below 4500 psi (31 025 kPa).

Since either port of the piston pump can be inlet or outlet, two replenishing and relief valves are required to provide relief protection.

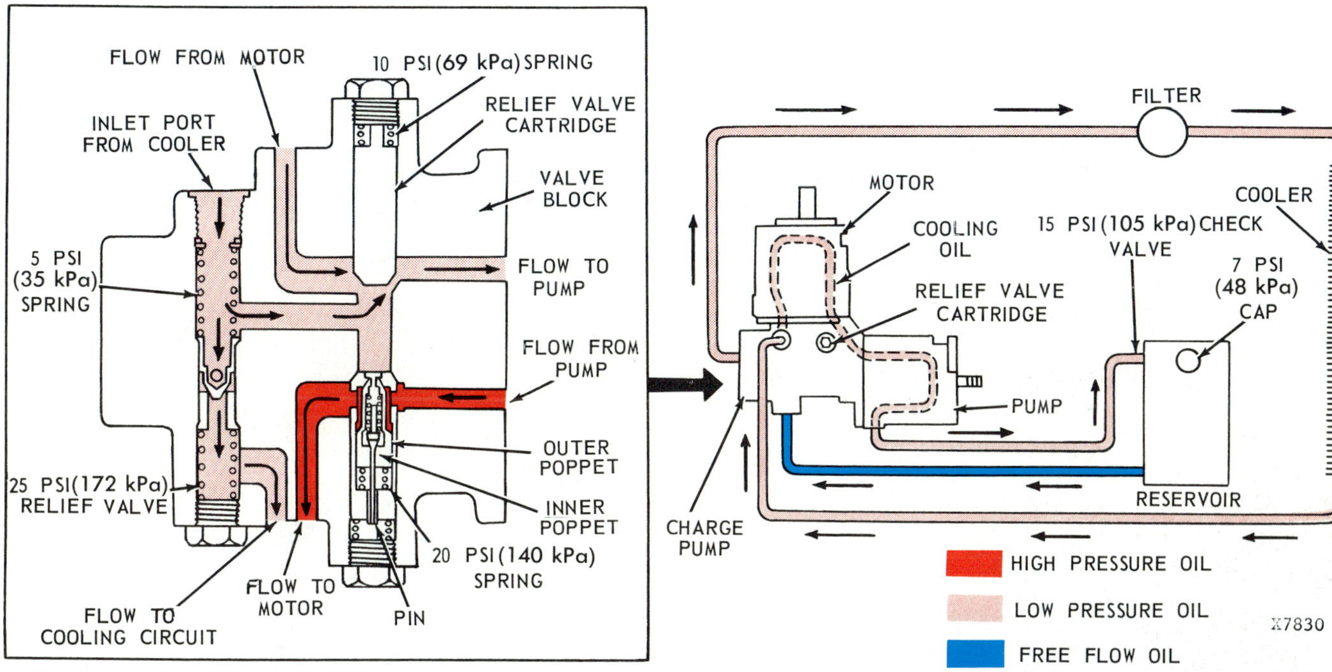

Fig. 17—Cooling Oil Flow in Hydrostatic Drive Axle

The pump for each axle is driven by the engine, while the motor for each axle drives that axle.

The machine is steered by actuating the pump swashplate on each axle. For example, a steering wheel can be linked indirectly to the swashplates for gradual turns. For sharp turns or reversing an axle, steering levers—one for each axle—can be used.

For reversing an axle, the pump swashplate is moved overcenter, reversing the flow circuit which drives the motor.

OTHER TYPES OF HYDROSTATIC DRIVES

A variety of hydrostatic units are available for driving machines from the center drive line or from the individual wheels.

To serve these needs, a wide variety of pump-motor locations are used.

We have looked at only a few of the more common designs.

ADVANTAGES OF HYDROSTATIC DRIVES

- *Variable speeds and torques*
- *Easy one-lever control*
- *Smooth shifting without "steps"*
- *Shifts "on the go"*
- *High torque available for starting up*
- *Flexible location—no drive lines*
- *Compact size for big power*
- *Eliminates clutches and large gear trains*
- *Reduces shock loads*
- *Low maintenance and service*

 Litho in U.S.A.

MAINTENANCE OF HYDROSTATIC DRIVES

Any hydraulic system is fairly easy to maintain: the fluid provides a lubricant and protects against overload. But like any other mechanism, it must be operated properly.

You can damage a hydraulic system by too much **speed,** too much **heat,** too much **pressure,** or too much **contamination.**

Keep the hydrostatic system clean. Impurities such as dirt, lint, and chaff, cause more damage than any one other thing. Seal any openings when doing service work to prevent dirt from entering the system. Be very careful that all parts are clean before replacing them.

NOTE: Any internal service on either the pump or motor must be done under extremely clean conditions. The tolerances in these units are similar to those in diesel injection pumps and equal care must be taken.

Before removing any part of the hydrostatic system for servicing, be sure that area is clean.

Use steam cleaning equipment, if it is available. However, do not let water enter the system. Make sure all hose and line connections are tight.

If steam cleaning is not possible, fuel oil or a suitable solvent may be used. Do not use paint thinner or acetone. Be certain to remove all loose dirt and foreign material that might get into the system when it is opened.

Use a clean work bench or table when disassembling the pump or motor for servicing. Do not perform internal service work on the shop floor or the ground or where there is danger of dust or dirt being blown into parts. Be sure all tools are clean and free of grease and dirt.

Have these items available before disassembling the pump or motor:

1. Several clean plastic plugs to seal openings when removing hydraulic lines or hoses.

2. Several clean plastic bags to place over open ends of lines or hoses. Secure bags with rubber bands.

3. A clean container of solvent to wash internal parts. Use clean compressed air to dry parts after washing.

Fig. 18—A Hydrostatic Drive Propels This Grain Combine

4. A clean container of transmission fluid to lubricate internal parts as they are assembled.

5. A clean container of petroleum jelly (petrolatum) to lubricate surfaces where noted during reassembly.

Whenever units are serviced, always install new O-rings, seal, and gaskets during reassembly. This will provide tight seals for mating parts and eliminate leakage.

Do not service the hydrostatic drive system with the machine engine operating unless specifically required.

Never operate the hydraulic system empty. Always check the oil supply after servicing the system.

If the machine is to be stored, be sure the hydraulic system is filled and the reservoir cap is tight. The oil will prevent rust and corrosion.

Always use clean oil in the system. Be careful to use the transmission fluid recommended in the machine operator's manual.

TESTING THE SYSTEM

The only testing required for many hydrostatic drives is for charging and operating oil pressures. Other systems may also require tests for oil flow rate and oil temperature.

Use the proper test equipment and procedures as given in the machine Technical Manual.

TROUBLESHOOTING OF HYDROSTATIC DRIVERS

Follow the general guide here for major problems and remedies. For specifics, see the machine Technical Manual.

NOTE: These troubleshooting charts include some items that will not apply to simpler drives.

MACHINE WILL NOT MOVE IN EITHER DIRECTION

Cause or Symptom	Remedy
System low on oil.	Check oil level.
	Check for oil leaks.
Faulty control linkage.	Check linkage for binding.
Disconnected coupling	Be sure coupling from engine to pump and from motor to gear train is installed correctly and are not slipping or broken.
Low or zero charge pressure.	Check charge pressure control valve. Also look for binding pump drive shaft or oil line restrictions.
Low on fluctuating charge pressure.	Check for air in system (if noisy). Also look for damaged charge pressure control valve (stuck open) or faulty check valves.

MACHINE MOVES IN ONE DIRECTION ONLY

Faulty control linkage.	Be certain control linkage is connected properly and not binding.
High pressure relief valve stuck open.	Replace valve which is stuck open.
One faulty check valve.	Replace both check valves.
Faulty directional control valve.	Check out speed control.

NEUTRAL HARD TO FIND

Cause or Symptom	Remedy
Faulty speed control linkage.	Adjust control linkage. Be certain linkage is not binding. Also, be sure displacement control valve is adjusted properly. Check out pump servo cylinders for proper adjustment.

SYSTEM OPERATING HOT

Oil level low.	Fill with oil.
Cooler clogged.	Clean cooler air passages.
Engine fan belt slipping or broken.	Tighten or replace fan belt.
Internal leaks. (Usually shown by loss of acceleration and power).	Check for stuck high pressure relief valve. Replace relief valve. Check for internal damage in pump or motor.

SYSTEM NOISY

Air in system.	Check oil supply. Be sure all air is out of system. Check for loose fittings or damaged lines and hoses.

HIGH LOSS OF OIL

Loose connections or leaking lines and hoses.	Tighten connections. Replace damaged lines and hoses. Check O-rings and seats.

ACCELERATION AND DECELERATION SLUGGISH

Air in system.	Remove all air from system. Check for low oil supply.
Control orifice partially blocked.	Check orifice and spool in displacement control unit for foreign material. If orifice and spool are clean, remove charge pump and clean passage between charge and displacement control unit with compressed air.
Internal damage in pump or motor.	Repair pump or motor as necessary.

Litho in U.S.A.

TRANSMISSION HARD TO SHIFT OR WILL NOT SHIFT

Cause or Symptom	Remedy
Speed control lever not in neutral.	Check and adjust linkage to be sure lever is in neutral.
Gearshift linkage out of adjustment.	Adjust gearshift linkage in all positions.

TEST YOURSELF

QUESTIONS

1. *(Fill in each blank with* ***high*** *or* ***low****.) Hydrostatic drives use fluids at ________ speeds but relatively ________ pressures.*

2. *How is a torque converter different from a hydrostatic drive as defined in Question No. 1?*

3. *What team of components is the heart of the hydrostatic drive?*

4. *Does a pump with a movable swashplate have a fixed or variable displacement?*

5. *What does the charge pump do for the hydrostatic drive oil circuit?*

(Answers on page 9 in back of text.)

TORQUE CONVERTERS / CHAPTER 6

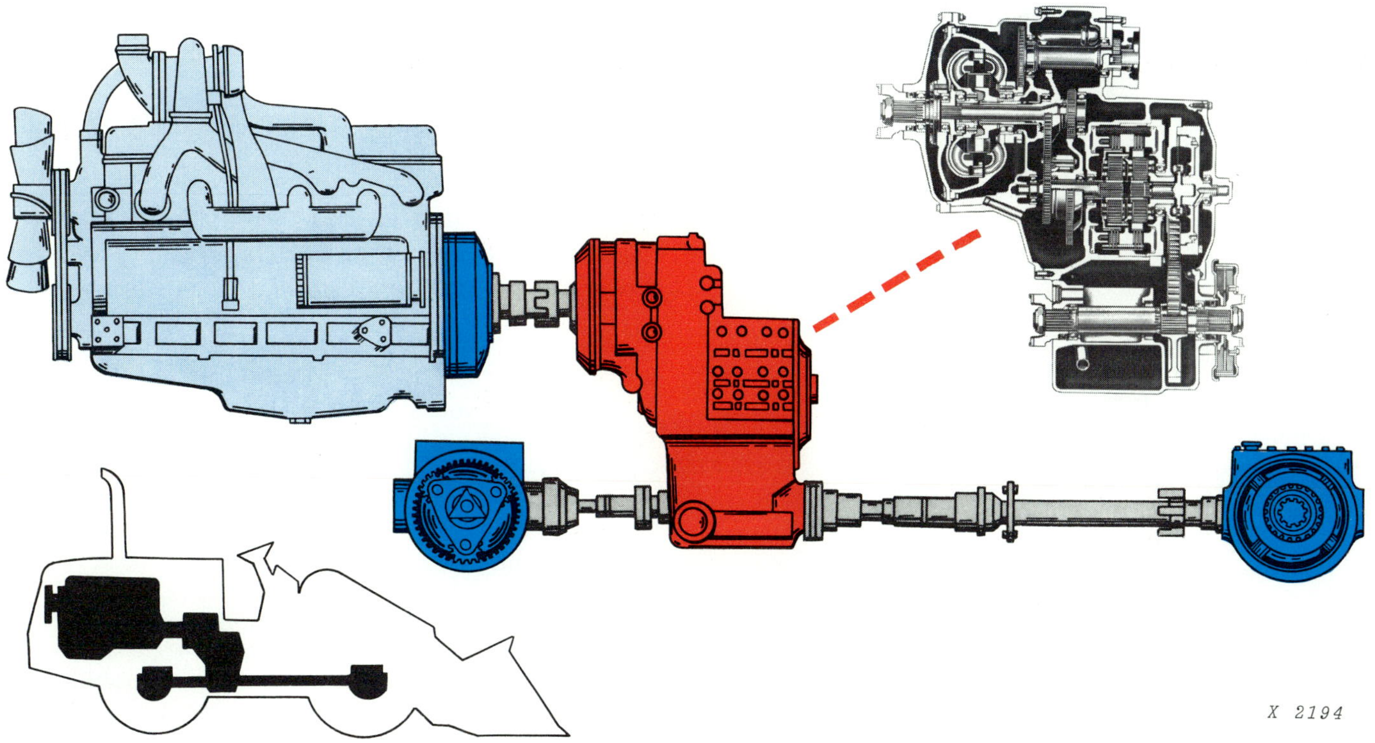

Fig. 1—Torque Converter In Complete Power Train

INTRODUCTION

A **torque converter** is an automatic *fluid* drive. It transmits engine torque by means of hydraulic force, shifting smoothly through an infinite number of speeds.

The automatic transmission of an automobile automatically shifts gears in response to torque requirements in addition to the automatic response of the torque converter which is a part of the automobile's automatic transmission system.

Actually a gear train is used with the torque converter to give extra speed ranges. But no gear train could give the infinite variations in speed and torque of a torque converter.

Acting as a *clutch,* the torque converter connects and disconnects power between the engine and the gear train. As a *transmission,* the converter gives many more speed ratios than are practical with a strictly mechanical gear box.

To compare a torque converter with a hydrostatic drive (Chapter 5), use this rule of thumb:

Hydrostatic drives are driven by fluids at *high pressure* but at relatively *low velocity,* while **torque converters** are driven by fluids at *low pressure* but at *high velocity.*

Here are the formulas:

- **Hydrostatic Drive** = *HIGH pressure + LOW velocity*
- **Torque Converter** = *LOW pressure + HIGH velocity*

HOW IT WORKS

To understand a torque converter, we must first look at a basic **fluid coupling.**

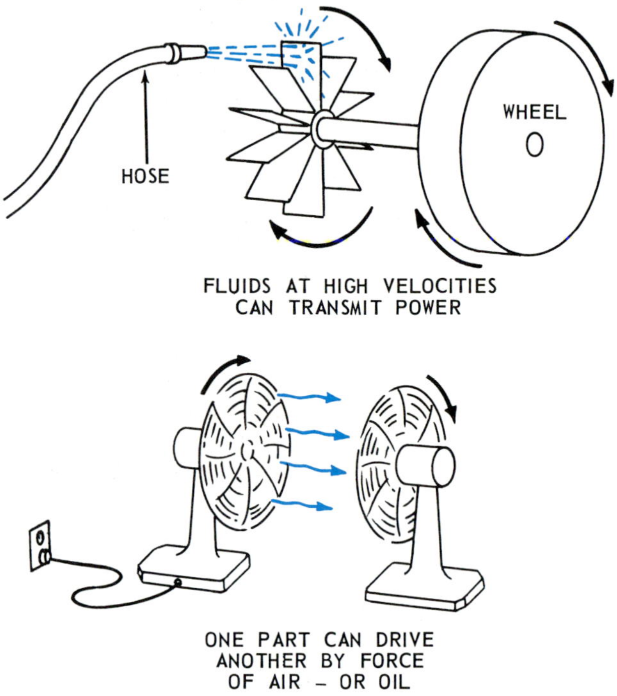

X 2195

Fig. 2—Basic Principles Of A Fluid Coupling

The basic principles of all fluid couplings are shown in Fig. 2.

At the top, a fluid at high velocity strikes a turbine and forces it to turn, driving the wheel. Thus, **torque is transmitted by a fluid.**

To change this torque, the velocity of the fluid is changed. At low velocity, the fluid will not even move the turbine. At higher velocity, the turbine starts turning and the wheel picks up speed.

This is something like putting two electric fans face to face as shown in the lower part of Fig. 2. By plugging in only one fan, we can cause the other one to rotate.

This principle is used in a fluid coupling as follows:

Inside an oil-filled housing (A, (Fig. 3) are two parts: the driving half, or **pump** (impeller), and the driven half, or **turbine**.

As the pump is turned by the engine, centrifugal force causes oil to be forced radially outward, crossing over and striking the vanes of the turbine. This rotates the turbine in the same direction and so couples the power.

Drawings B, C, and D explain how the flow of oil drives the turbine.

In B, fluid is placed in a bowl and lies level.

In C, the bowl is spun rapidly and centrifugal force causes the fluid to climb up and spill over the outside edge of the bowl.

In D, another bowl is placed down over the first one. Now when the bowls are spun an axial flow or circuit is created and turning force is transmitted between the driving bowl and the driven bowl.

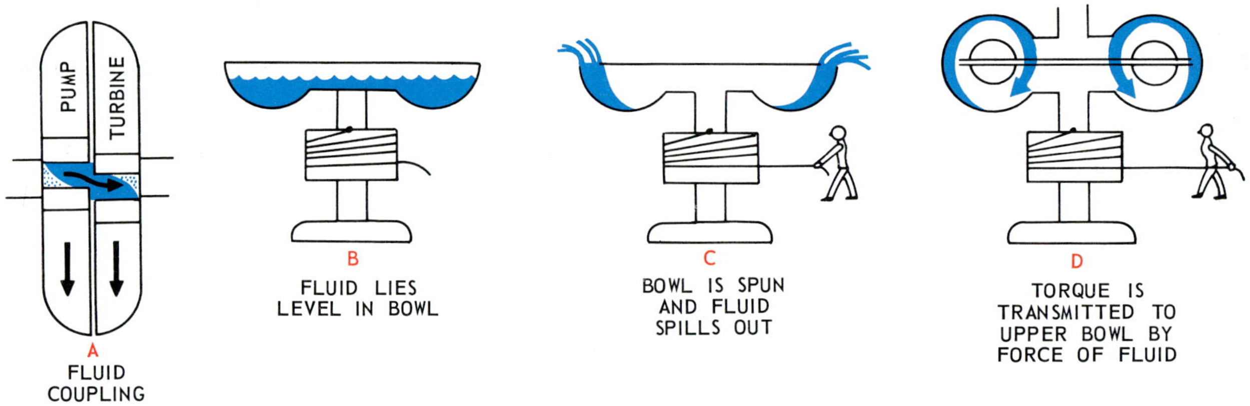

X 2196

Fig. 3—Operation Of A Fluid Coupling

Litho in U.S.A.

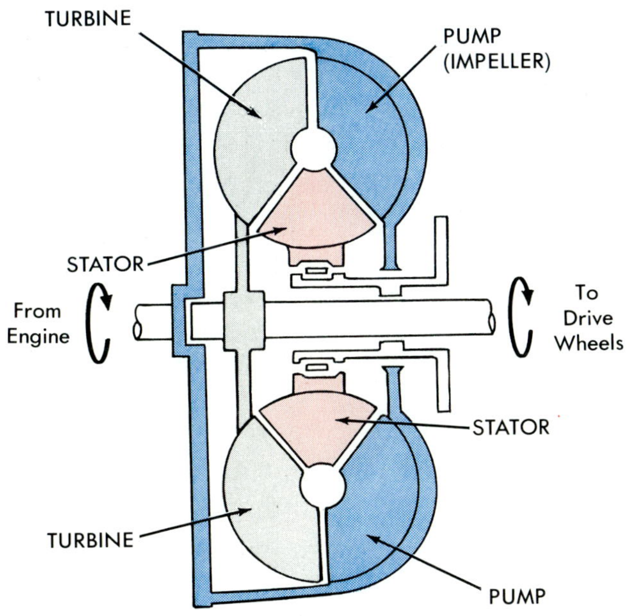

Fig. 4—Torque Converter

Torque is thus transmitted, but it is not increased.

This is where the **torque converter** goes beyond the basic fluid coupling, for the converter can **multiply torque.**

A torque converter (Fig. 4) looks much like the fluid coupling we have just described. The main difference is that the torque converter has—in addition to the driving pump and the driven turbine—a set of blades or vanes called a **stator.**

The stator vanes change the direction of oil flow after it has gone through the turbine and sends it back to the pump. This enables the pump to increase the twisting force or *multiply* the torque.

Since the converter is a closed unit, this flow is repeated continuously. Many streams of fluid act against many vanes at once and this is what gives the power to drive a heavy machine.

OIL FLOW IN CONVERTER

Let's look at the flow of oil in the converter during two cycles:

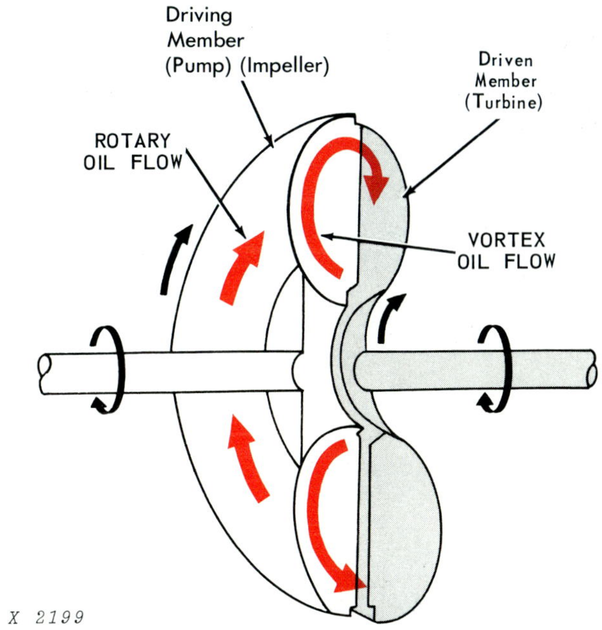

Fig. 5—Vortex And Rotary Flow In A Fluid Coupling

1) Increasing the torque

2) Decreasing the torque

INCREASING THE TORQUE

Remember that the pump is driven by the engine, while the turbine receives fluid energy from the pump and sends it to the drive wheels.

Also remember how centrifugal force sets up a continuous circular flow in the coupling (Fig. 5).

This circular flow of oil *between* the pump and turbine is called **vortex flow.**

Another flow is set up *around* the pump and turbine to form a coupling; this is called a **rotary flow.**

The action of these combined oil flows will transmit torque *but not increase it.*

Increasing the torque is where the stator comes into play.

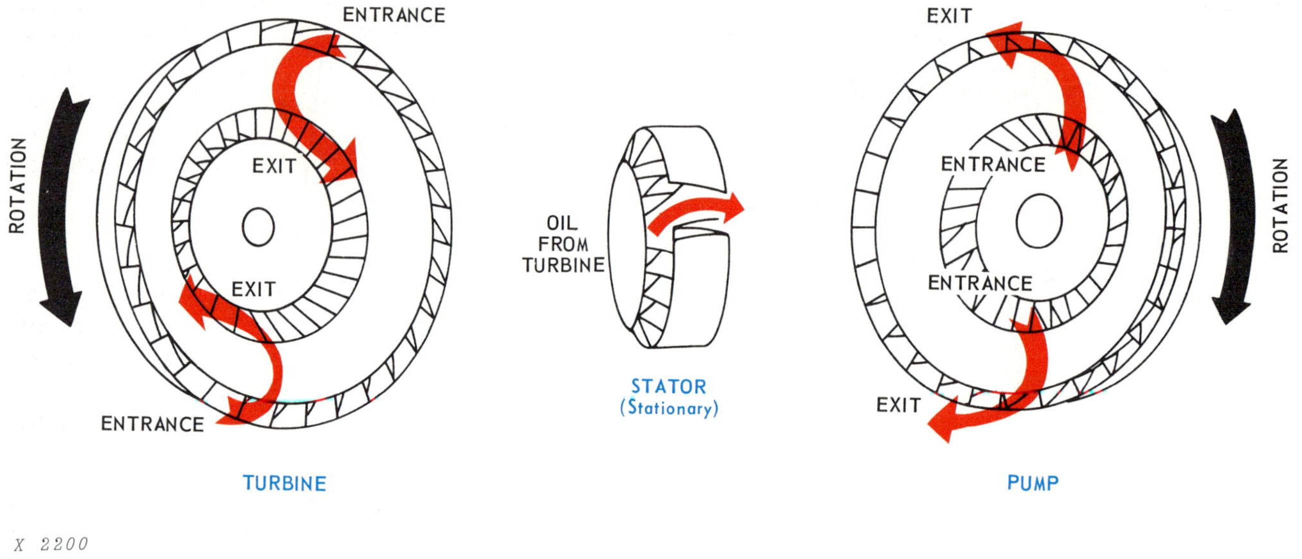

Fig. 6—Oil Flow Through Pump, Turbine, And Stator

Fig. 6 shows how oil emerges from the turbine in reverse compared to pump rotation. Unless this oil flow is turned around, it will cause a loss of power.

Note in Fig. 6 that the oil passages at the rim of the turbine where the oil enters must become smaller as they approach the smaller diameter of the turbine. As the same volume of oil must squeeze through these funnel-like passages, the oil stream will speed up when it leaves the turbine. This speed is used to increase torque by directing it against the **stator,** which acts as a fluid lever or fulcrum. The stator changes the direction of flow and sends the oil into the pump in the same direction as pump rotation.

Let's see how the stator does its job (Fig. 7). A stream of oil aimed at a flat surface (A) splashes off at various angles. The oil can be made to flow more smoothly by curving the entrance (B) and can be reversed by more curving (C) with a resulting increase in force as indicated by the large arrow.

The stator has curved blades (as in C) which the oil strikes as it leaves the turbine. These blades turn the oil back in the direction of pump rotation.

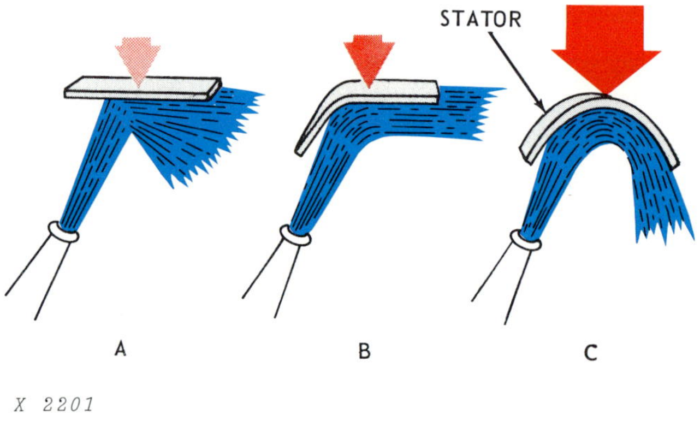

Fig. 7—Stator As A Fluid Lever

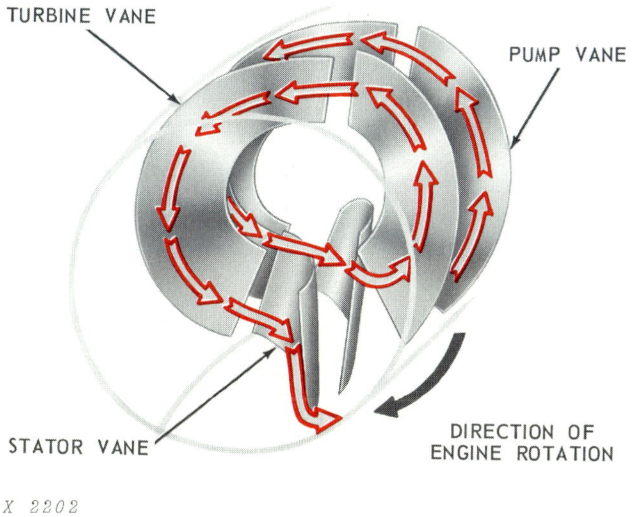

Fig. 8—Flow Direction Is Reversed By The Stator Vanes

Now that the stream of oil is moving in the same direction but at a greater velocity, it enters the pump smoothly (Fig. 8). Its velocity is added to that developed in the pump so that the total velocity at the pump exits has been increased.

This regenerating action is the key to multiplying torque in a torque converter.

To change the direction of oil flow, the stator must be stationary during the increase in torque. However, once the pump and turbine are turning at the same speed it would create resistance. Therefore, the stator is sometimes mounted on a freewheel clutch so that it can turn in one direction only (once torque stops increasing). (In other torque converters, the stator may be fixed to the converter case.)

DECREASING THE TORQUE

Torque is increased as long as the engine is accelerating to get the machine under way. But as the engine speeds up, the turbine also speeds up, which causes the vortex flow of oil in the converter to decrease. At the same time, the rotary flow increases.

Vortex flow keeps on changing to *rotary* flow (Fig. 9) until the pump and turbine are "locked up" and all torque increase stops.

The torque converter now acts as a simple fluid coupling, sending the same torque it receives on to the drive wheels.

The torque converter is able to automatically reduce or increase torque in infinitely small steps to match the needs of the machine and its driver. This has the same effect as shifting speeds in a gear transmission except that it is done smoothly and automatically while "on the go."

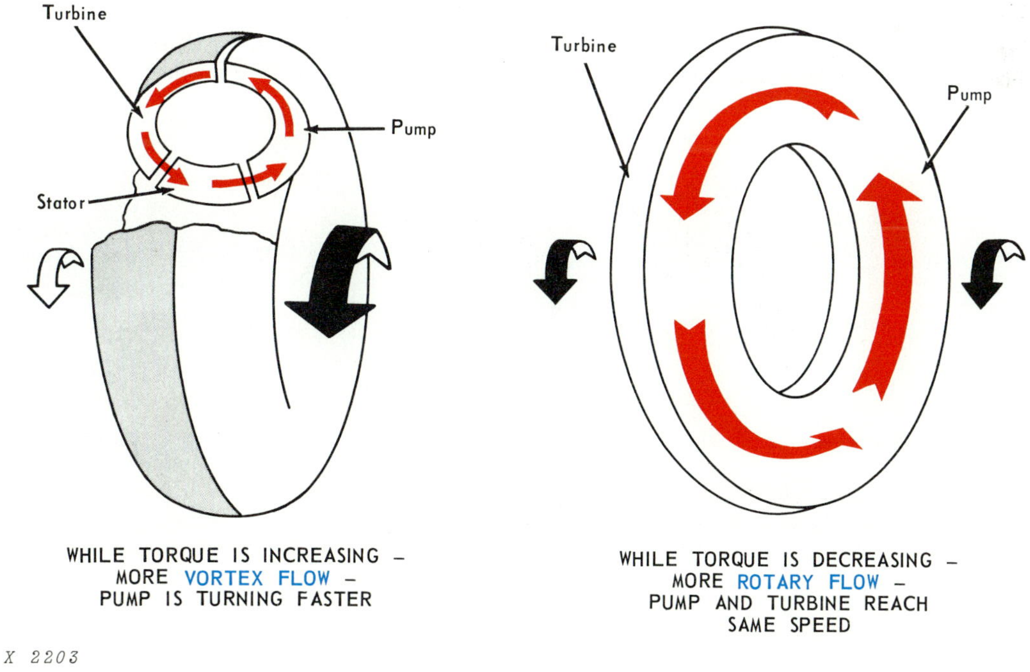

Fig. 9—How Vortex Flow Changes To Rotary Flow As Torque Is Reduced

VARIATIONS ON TORQUE CONVERTERS

The chart below shows some of the variation in the number of elements used in several torque converters.

Torque Converter Elements	*Design A*	*Design B*	*Design C*	*Design D*
Pump	*2*	*1*	*1*	*1*
Stator	*2*	*1*	*2*	*1*
Turbine	*1*	*2*	*1*	*1*

The design of a torque converter must match the engine torque and road speed for each application. In off-the-road machinery, units are matched to horsepowers (kilowatts) ranging from 40 to 600 (30-450). But the basic principles we've covered apply to all.

TORQUE CONVERTER TRANSMISSIONS

The torque converter is but one part of the complete transmission (Fig. 10).

Here are the major components:

- **Converter Section**
- **Range Gear Section**
- **Final Drive Section**
- **Hydraulic Control System**

Let's look at each part and see what is does for the complete transmission.

CONVERTER SECTION

We have looked at torque converters with three elements—one pump, one turbine, and one stator.

Now we will examine a *twin-turbine* model (Fig. 11) which has one pump and one stator, but *two* turbines (first and second).

The first turbine is shown in blue, while the second is shown in red.

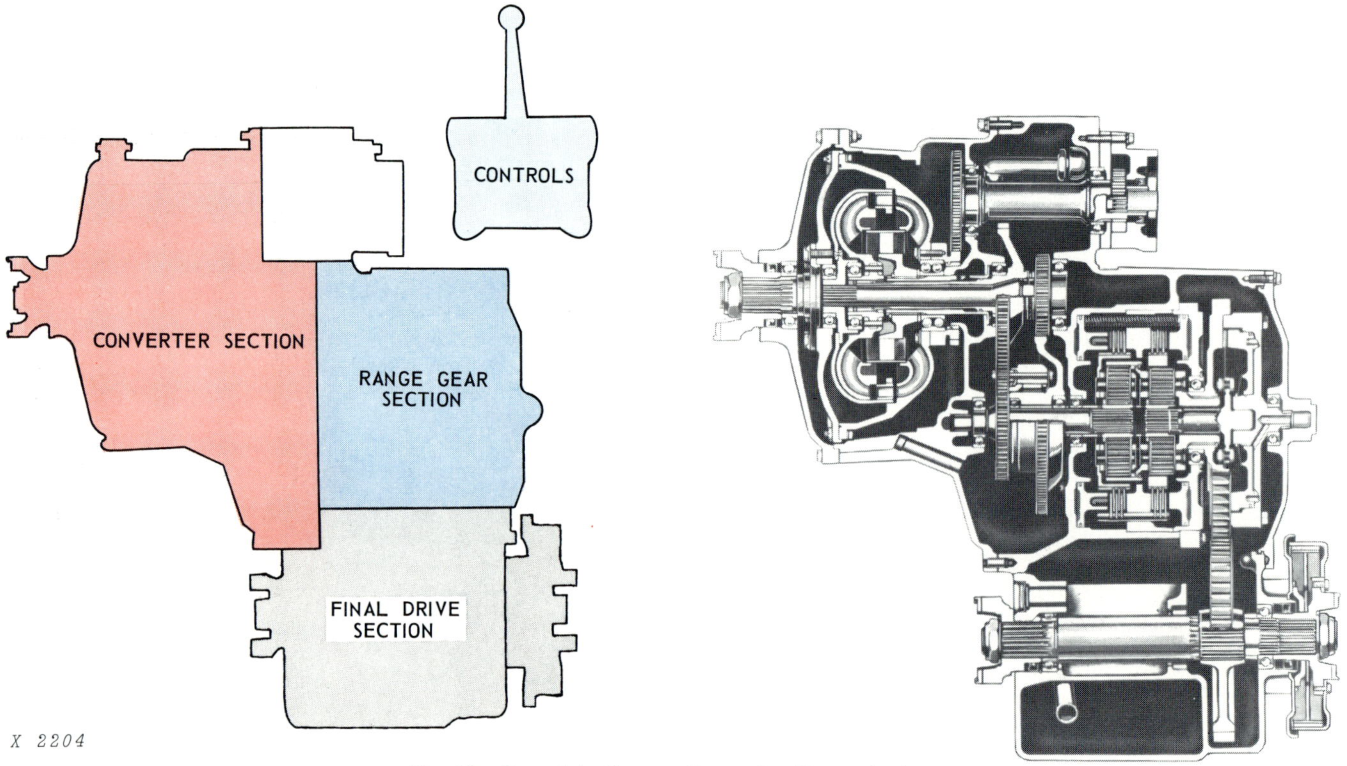

Fig. 10—Complete Torque Converter Transmission

Each turbine is connected to its individual output gear set. In reality, two converters have been combined into one. The first (blue) turbine is connected to its output shaft by a freewheel clutch.

Here is how the two turbines work together:

When torque demand is high, the freewheel clutch is engaged and the first turbine, assisted by the second turbine, drives the gears. When the machine speeds up, torque demand drops. Then the second turbine takes over the entire load and the freewheel clutch disengages the first turbine.

As a result, the first turbine provides *high torque and low speed,* (for starting up and loading) while the second turbine provides *higher speed with lower torque* (for travel).

A combining gear set directs torque from the first and second turbine (or the second turbine only) to the range gear section.

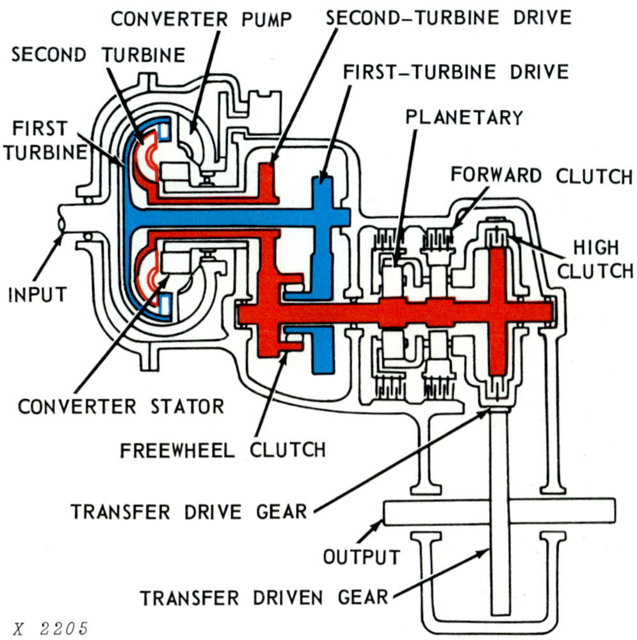

Fig. 11—Twin-Turbine Torque Converter

RANGE GEAR SECTION

Since torque is reduced and increased automatically in the torque converter, only a few gear sets are normally required in the transmission.

However, as the torque converter rotates in only one direction, it is necessary to have a reverse gear. In some applications, it is also desirable to have low and high range gears as shown in Fig. 11.

Simple planetary gear sets meet the needs of extra gear ranges and lends themselves to hydraulic control.

NOTE: Planetary gear sets and their control by hydraulic clutches are explained in Chapter 4.

FINAL DRIVE SECTION

The final drive section includes the transfer drive gear, transfer driven gear and output shaft (Fig. 11). The output shaft provides for one or for two outputs from the same common shaft. Two outputs can be used to propel a four-wheel drive machine as shown in Figs. 11 and 15.

Note also that by adding the transfer gear and one clutch (high range) in Fig. 11, another forward speed range can be obtained.

HYDRAULIC CONTROL SYSTEM

The hydraulic control system uses oil to do these jobs:

- **Oil flow lubricates and cools the parts**
- **Oil pressure engages the clutches**
- **Oil velocity drives the turbines**

Let's use the hydraulic controls for the twin-turbine converter we have just described to see how a typical system works.

There are four basic circuits as shown in Fig. 12:

1) Oil Pump and Filter Circuit (shown by blue lines).

2) Main-Pressure Regulator Valve and Converter-In Circuit (shown by red lines).

3) Converter-Out, Cooler, and Lubrication Circuit (shown by dotted blue lines).

4) Selector Control Valve Circuit (shown by dotted red lines).

Let's build up the system and explain each circuit.

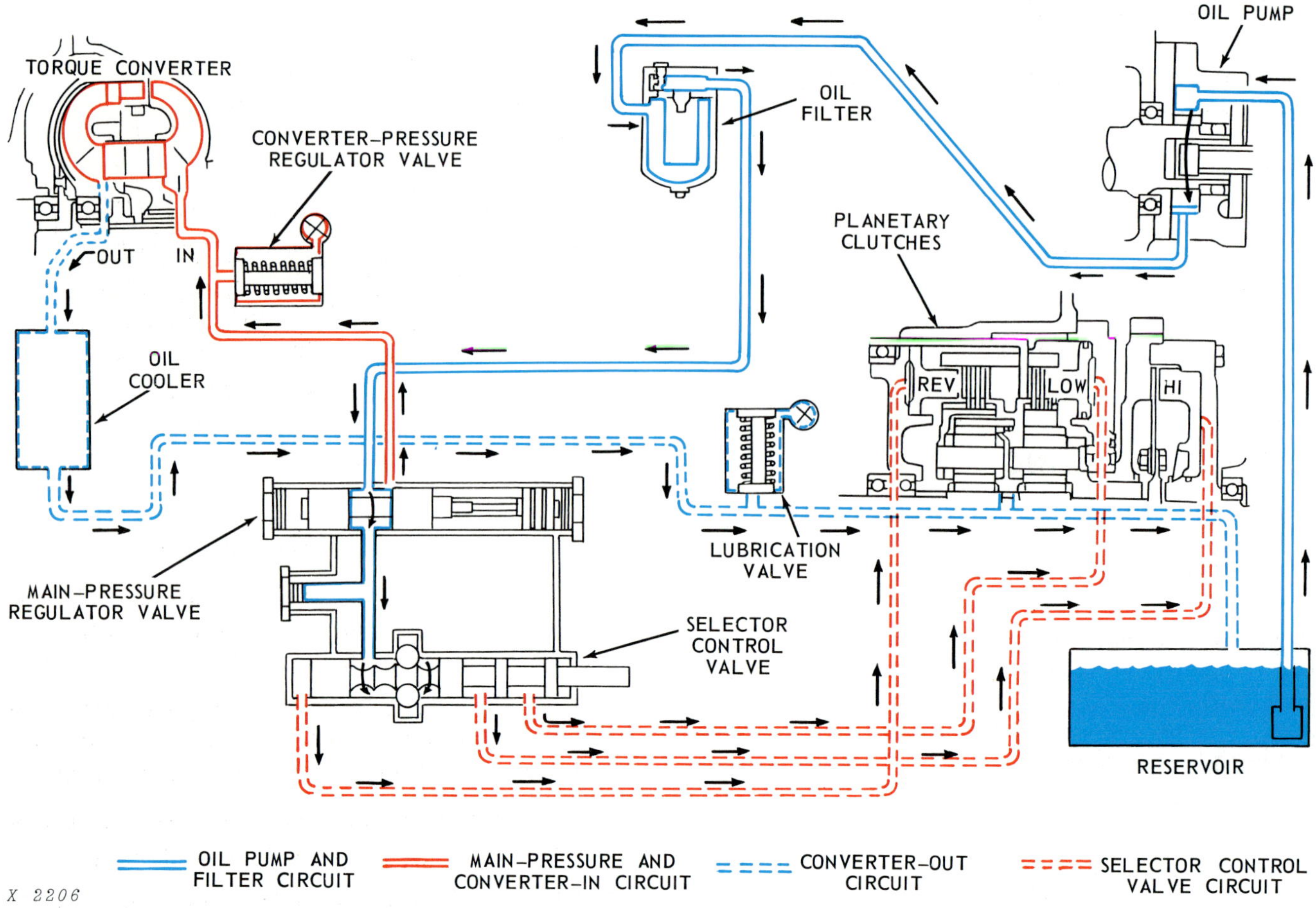

Fig. 12—Hydraulic Control System For Torque Converter Transmission

Oil Pump And Filter Circuit

Oil is drawn from the transmission reservoir by the oil pump as shown in Fig. 12. The pump delivers its entire output to a full-flow oil filter for cleaning. From the oil filter, the oil supply is sent to the main-pressure circuit.

Main Pressure Regulator Valve and Converter-In Circuit

The main pressure regulator valve provides pressure for the planetary clutch packs, directs oil to the selector control valve, and supplies pressure oil into the torque converter. A converter pressure regulator valve in the converter-in line limits the oil pressure there.

Converter-Out, Cooler, and Lubrication Circuit

The torque converter is continuously filled with oil during operation. Rotation of the converter pump imparts energy to the oil which, in turn, drives the turbines. The oil then flows between the stator vanes which redirect it to the pump.

Oil flowing out of the converter is directed into the oil cooler as shown in Fig. 12. The cooler is a heat exchanger in which the oil flows through water- or air-cooled passages.

From the cooler, oil flows to all passages and outlets in the lubrication circuit (dotted blue lines). A lubrication valve between the cooler and lubrication system returns all excess oil to the transmission reservoir.

Selector Valve Circuit

Pressure oil from the main-pressure regulator valve flows into the selector valve bore and surrounds the valve in the area of the detent notches. From this area, main pressure oil is available for operating the low, high, and reverse range planetary clutches (see dotted red lines).

Moving the selector valve allows oil to charge the selected clutch line and to engage that clutch.

This completes the four basic control circuits in our torque converter transmission.

NOTE: For details on hydraulic components, see the FOS "Hydraulics" manual.

SUMMARY: FEATURES OF TORQUE CONVERTERS

1. *Multiply torque*
2. *Provide infinite speed ranges*
3. *Shift smoothly and automatically*
4. *Cushion shock loads on drive lines*
5. *Help to dampen vibrations*

TROUBLESHOOTING

INTRODUCTION

Oil circulates at high velocity within the torque converter and any foreign material it carries will rapidly wear down the edges and pit the turbine vanes (Fig. 13), changing their effective shape.

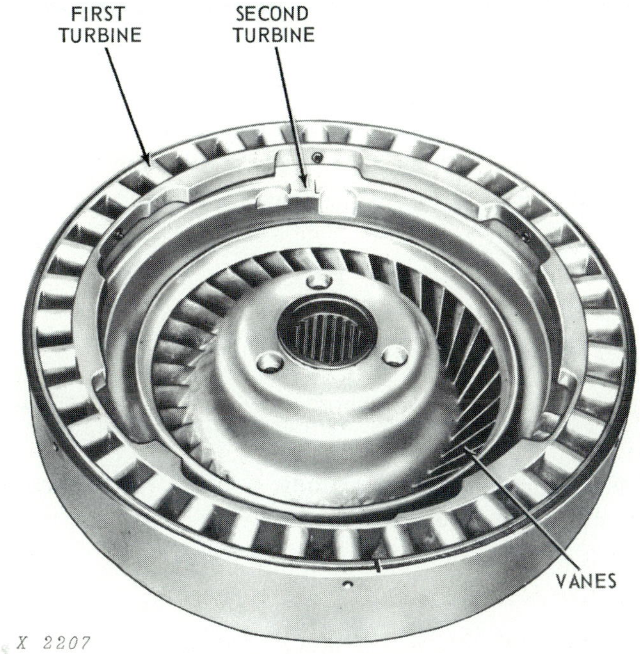

Fig. 13—Turbines For Twin-Turbine Torque Converter

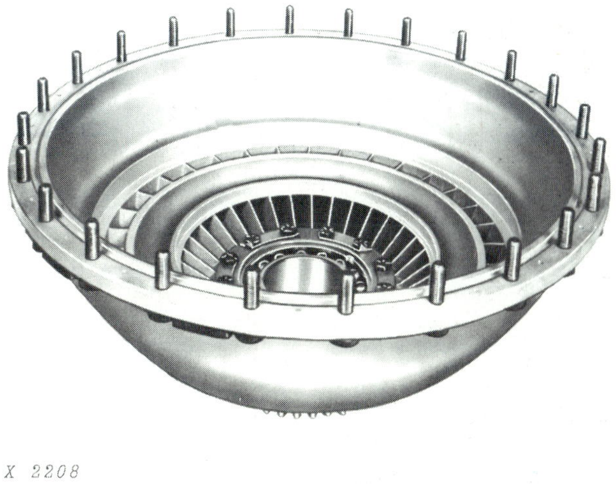

Fig. 14—Pump For Torque Converter

Vane damage will also cause the turbines to become unbalanced. In addition, dirty oil will damage bearings and seals.

Some torque converters contain parts made of lightweight aluminum alloys (Fig. 14). Converter housings are usually made of cast aluminum.

Be sure to handle all converter parts carefully to prevent nicking, scratching, and denting.

Parts which fit together closely but with operating clearance will stick if damaged only slightly. Parts which depend upon smooth surfaces for sealing may leak if scratched.

All these parts should be carefully handled and protected during removal, cleaning, inspection, and installation.

Use these rules to help prevent failures:

1. *Be sure the oil is kept clean.*
2. *Service the system at proper intervals.*
3. *Repair it only if you are qualified.*
4. *Use all special tools recommended.*
5. *See the machine Technical Manual for details.*

PRELIMINARY INSPECTION

The operator and serviceman can use his eyes, ears, and even his nose to head off trouble before serious damage is done. Here's how:

Look *Check oil levels.* Refer to the operator's manual.

Check gauges. When starting, running or stopping, watch the gauges; they tell the story in terms of pressure and temperature.

Inspect for leaks. Leaks, crimped oil lines, and clogged filters — all affect the converter output.

Examine the oil. Look for water, dirt, and particles from the converter and the clutch plates.

Listen *Unusual noises.* Listen for squealing from a stuck valve, or grinding or grating sounds from inside the converter.

Smell *Overheating.* A strong odor of overheated oil is a major trouble sign. Find the cause at once.

Because one trouble can have the same symptoms as another, only good instruments in the hands of trained servicemen can detect the difference.

Once trouble has been detected in the field, check out the unit with the proper testing tools.

TROUBLESHOOTING PROBLEMS

The troubleshooting given here will cover four common areas of trouble:

1. *Overheating*
2. *Noise*
3. *Leaks*
4. *Machine Response*

Let's examine each kind of trouble.

1. Overheating

Overheating is a major problem in converter operation. It is affected by the design, the type of work, the operator, the air temperature, and the condition of the unit.

Overheating can cause a loss of power and can damage seals and gaskets and warp metal parts.

A converter may overheat if the *work is heavy*—not always, but heavier work will generally mean more heat.

If the converter is *undersized* for the normal work of the machine, it will operate at low efficiency and will tend to overheat. The operator can usually relieve the load on the converter by operating in a lower gear, or by reducing the load.

Whenever you receive complaints that a converter in a machine is overheating, try to find out by watching or by questions whether the correct gears and work methods are being used.

Air in the converter will also cause overheating. Torque converters can work properly only if they are filled with fluid. Air mixed with the oil will cause poor performance, overheating, and possibly serious damage.

Air may enter the system:

(1) If the fluid in the reserve tank (if used) is low enough to permit the charging pump to suck air.

(2) If a moderately low level of oil permits sucking of air while the machine is working on steep slopes.

(3) If there is a leak in the suction line pump gaskets or O-rings (suction leaks may be too small to show up by outside leaking of oil).

(4) If the oil or filters are changed or when the lines are opened for any reason.

SUMMARY: OVERHEATING

Let's recap the major causes of overheating in the torque converter.

First, overheating is not only a major problem, it is also a major symptom.

Although normal operating temperatures can be exceeded very rapidly, a machine using a torque converter matched to the job should not exceed its normal temperature when properly used.

Second, the cause of overheating may be found in one or more areas outside of the converter.

Before deciding that the converter is the cause of overheating, check these possibilities:

(a) Air in fluid system.

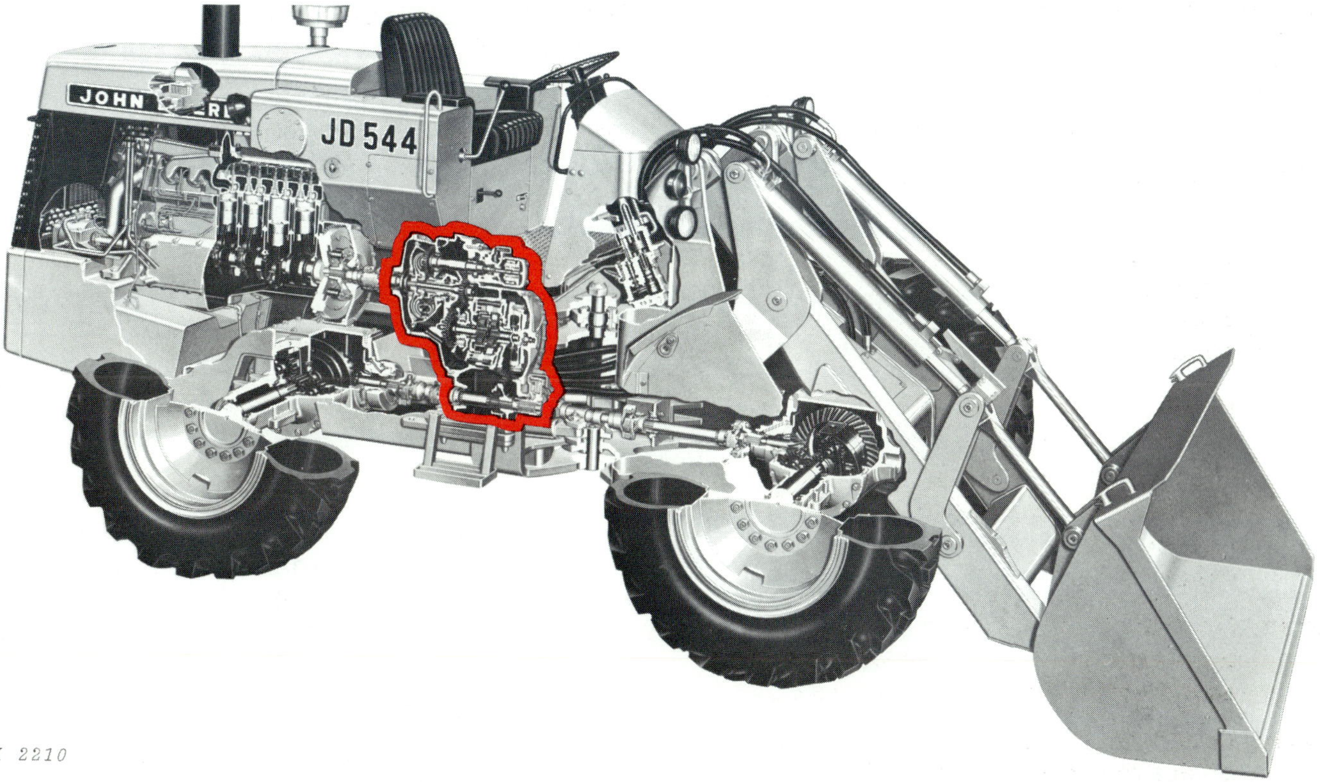

X 2210

Fig. 15—Torque Converter Transmission In Four-Wheel Drive Machine

(b) Lack of cooling—plugged cooler core, low coolant level, defective water pump (water-cooled systems only).

(c) Low fluid level—clogged filter, excessive leakage past converter seals, restricted oil line, defective oil pump.

(d) Slipping clutches in planetary gear sets.

Third, failure of the converter may cause the overheating. Worn edges and pits on vanes of pump, stator, or turbines will reduce their efficiency.

2. Noise

Unlike overheating, which can be tested, noise is hard to explain to another person.

A new operator or serviceman may never hear a noise directly related to a converter failure.

Yet, an unusual noise heard by an experienced operator or serviceman may be the first sign of damage in the converter.

Noise from a converter malfunction may be a whining or growling and may be steady or intermittent.

Worn or dry bearings often produce a hissing noise that will develop into a bumping or thudding sound when they completely fail.

Other sources of noise are: Worn gears, worn or bent shafts, excessive shaft endplay, shafts misaligned with the engine, and worn freewheel clutches. All these noises mean a possible converter failure.

A mechanic's stethoscope is a valuable aid for isolating noises in the converter.

3. Leaks

Converter leaks can be one of two types:

- *Internal leaks*
- *External leaks*

 Litho in U.S.A.

INTERNAL LEAKS

By internal leaks we mean those *within* the converter.

As we learned earlier, the converter uses large amounts of oil at high velocity.

If oil is lost from the converter housing by leaks in the pump, turbine, or stator, a loss of power or erratic operation will occur.

Leaks may be caused by the wrong torque on converter bolts.

On some converters, the housing which covers the converter can be removed to determine if a leak has occurred in the converter. Check by starting the engine and operating the transmission until the oil leak shows up.

If leaks form around the converter cover, check the tightness of the cover bolts with a torque wrench. If this fails to correct the leak, disassemble the cover, check the machine surfaces of the cover and flywheel, and install a new gasket.

EXTERNAL LEAKS

By external leaks we mean those occurring outside the converter but still affecting its operation.

These include leaks at cooler lines, filter lines, and pressure or temperature gauge fittings.

Visually check all fittings and oil lines for leakage.

4. Machine Response

Normally a malfunction in the converter will affect the machine's response to load and speed changes.

A machine which lacks power and acceleration at low speed may have a turbine freewheel clutch failure.

Changes in hydraulic pressure, flow, and temperature also affect the performance of the converter, and thus affect the machine's performance.

If you picture the converter as in Fig. 7 (a flow of oil from a nozzle) it is easy to see how heavy cold oil, low pressure, or low flow will affect a response.

TESTING THE TORQUE CONVERTER

As in trouble shooting, *testing* is most effective when the engine, converter, and gear train are regarded as a unit, one part affecting the other.

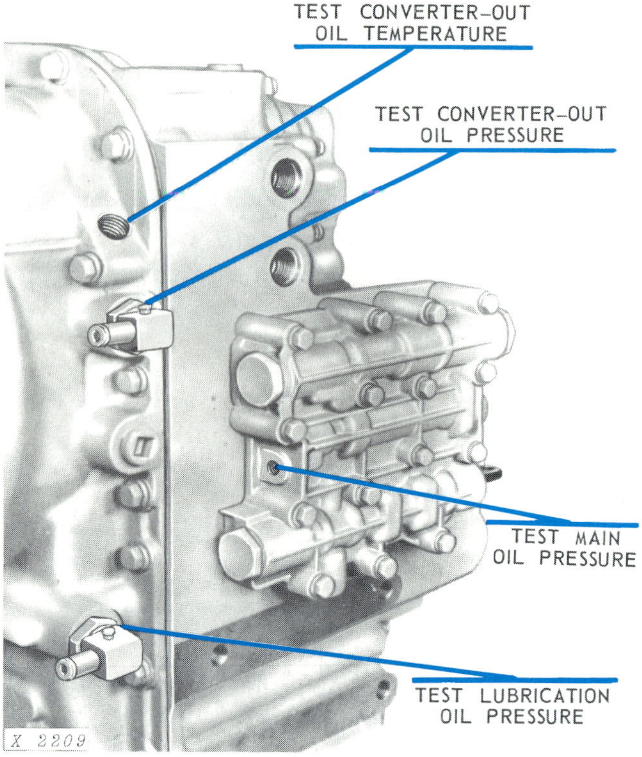

Fig. 16—Pressure And Temperature Check Points On A Typical Torque Converter Transmission

Fig. 16 shows the points where pressures and temperature may be tested on a *typical* torque converter.

Before testing, several checks should be made.

BEFORE TESTING

1. Check the fluid level. Be sure it is not *above* or *below* the required level.

2. Start the engine and warm the engine and transmission to operating temperature.

3. Shift into each gear and operate for a minimum of 15 seconds, checking the selector valve detent positions against the related positions on the shift indicator (if equipped).

4. *Never allow the transmission to heat up beyond the maximum operating temperature.*

TESTING PRESSURE AND TEMPERATURE

Attach the necessary temperature and pressure gauges to the converter. Be sure that all tests are performed as required by the machine Technical Manual.

For example, tests made on the typical unit in Fig. 16 are as follows:

(1) Main Pressure (at full throttle—no load).

(2) Converter-Out Pressure (at full throttle—no load and at full throttle stall in high range).

(3) Lubrication Pressure (at full throttle—no load).

(4) Converter-Out Temperature (during normal operating conditions).

TRANSMISSION STALL TEST

IMPORTANT: Be sure to consult the machine Technical Manual before conducting a stall test. Some manufacturers do not recommend this test.

The stall test tells whether or not the engine, torque converter, and transmission are performing satisfactorily *as a unit.*

The test is made while the output shaft is braked and the engine is running at full throttle.

Refer to the engine-converter matched performance curve to find the acceptable speed at stall (see the Technical Manual).

Procedure

1. With the transmission warmed up, connect an *accurate* tachometer to the engine.

2. Block the machine securely and shift to the desired range.

3. Accelerate the engine to full throttle.

4. After reaching a stabilized engine speed, record the engine speed.

IMPORTANT: Because of rapid heating of the oil, never hold the machine stalled for more than a few seconds at one time.

Results

If the engine and transmission are failing, it will show up when the actual engine speed in the stall test is compared to the normal speed in the engine-converter chart.

Allow for variations in the test speed due to climate, altitude, engine accessory loads, and power input at converter, all of which affect the stall test results.

TEST YOURSELF

QUESTIONS

1. (Fill in the blanks.) *A ________ ________ can only transmit torque, while a ________ ________ can multiply the torque it receives.*

2. *What part in a torque converter directs the fluid back to the pump?*

3. *(Fill in the blanks.) Once torque demand is reduced in the converter, the pump and turbine "lock up" and ______ flow is changed to ______ flow.*

4. *True or false? "The stator will turn in either direction."*

5. *True or false? "Only one turbine can be used in a single torque converter."*

6. *Match each item below with the best match at the right.*

a. Oil flow	*1. Drives the turbines*
b. Oil velocity	*2. Engages the planetary clutches*
c. Oil pressure	*3. Lubricates and cools*

(Answers on page 9 in back of text.)

DIFFERENTIALS / CHAPTER 7

Fig. 1—Differential In Complete Power Train

INTRODUCTION

The *differential* does two jobs:

- **Transmits power "around the corner" to the drive axles.**
- **Allows each drive wheel to rotate at a different speed and still propel its own load.**

The ring gear and bevel gears direct the power to the axles, while the bevel pinions give the differential action (Fig. 2).

We saw in Chapter 1 how the differential works. Let's review it briefly here.

OPERATION

When the machine is moving straight ahead, *both wheels are free to rotate* as shown in Fig. 2.

Engine power comes in on the pinion gear and rotates the ring gear. The four bevel pinions and the two bevel gears are carried around by the ring gear and all gears rotate as one unit. Each axle receives the same rotation and so each wheel turns at the same speed.

When the machine turns a sharp corner, only *one wheel is free to rotate* as shown in Fig. 2.

Again engine power comes in on the pinion gear and rotates the ring gear, carrying the bevel pinions around with it. However, the right-hand axle is held stationary and so the bevel pinions are forced to rotate on their own axis and "walk around" the right-hand bevel gear.

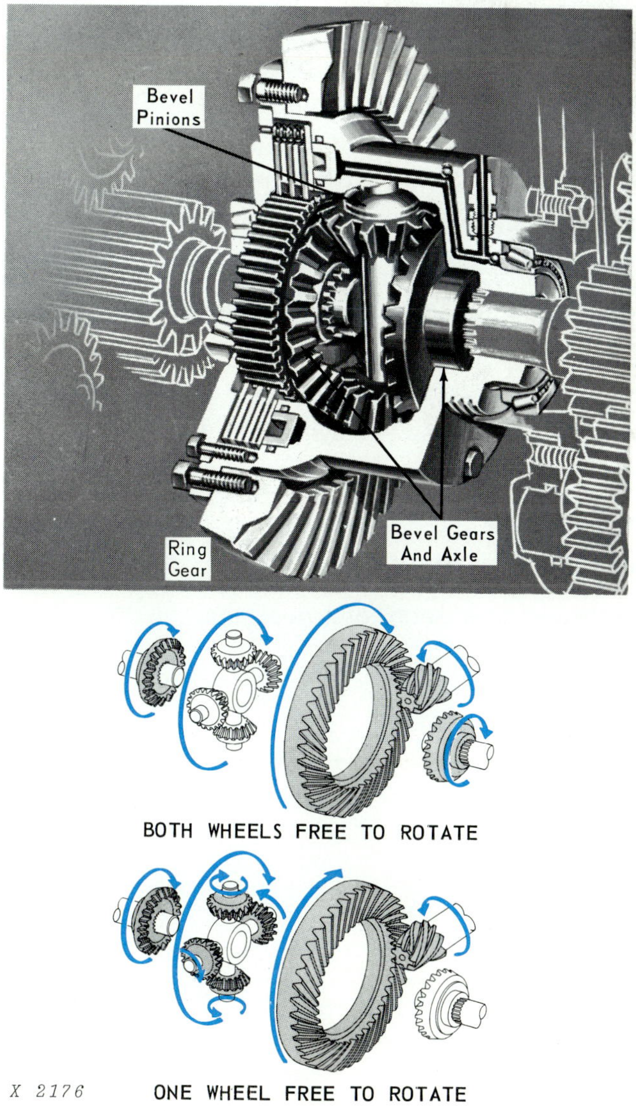

Fig. 2—Differential In Operation

Since the bevel pinions are in constant mesh with both bevel gears, the left-hand bevel gear is forced to rotate because it is subjected to the turning force of the ring gear which is transmitted through the bevel pinions.

During one revolution of the ring gear, the left-hand gear makes *two* revolutions—one with the ring gear and another as the bevel pinions "walk around" the other bevel gear.

As a result, when the drive wheels have unequal resistance applied to them, the wheel with the least resistance turns more revolutions. As one wheel turns faster, the other turns slower by the same amount. However, both wheels still propel their own loads—but at different speeds.

Differential action can be a disadvantage when the wheels slip. For example, when a farm tractor is plowing, one wheel may lose resistance on slick ground and start spinning while the other wheel holds. Then the driving power is limited by the "wasted" power sent to the wheel that slips.

To prevent this loss of power, *differential locks* are often used, as we'll see now.

DIFFERENTIAL LOCKS

A *differential lock* directs power equally to both wheels by locking out the differential. This prevents the usual loss of traction when one wheel is slipping.

Three types of differential locks are used:

- **Mechanical**
- **Hydraulic**
- **Automatic (No-Spin)**

Let's see how each lock operates.

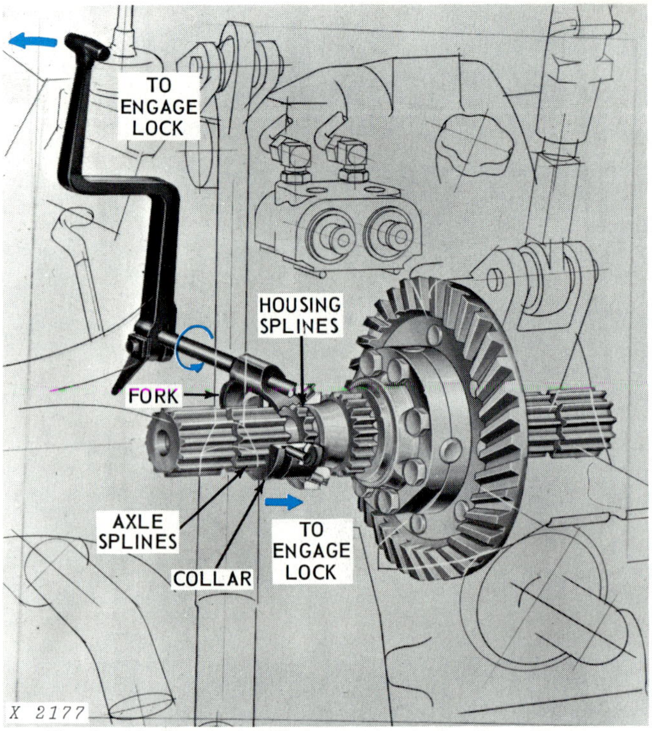

Fig. 3—Mechanical Differential Lock On Farm Tractor

MECHANICAL LOCKS

The simplest differential lock is the *mechanical* one shown in Fig. 3.

When the operator moves the upper lever, the fork rotates and slides the collar to the right. This engages the differential housing splines to the collar. Since the collar is already on the axle splines, both the axle and the housing are forced to rotate as a unit. This in turn prevents the bevel gear on the axle from turning and so locks the differential.

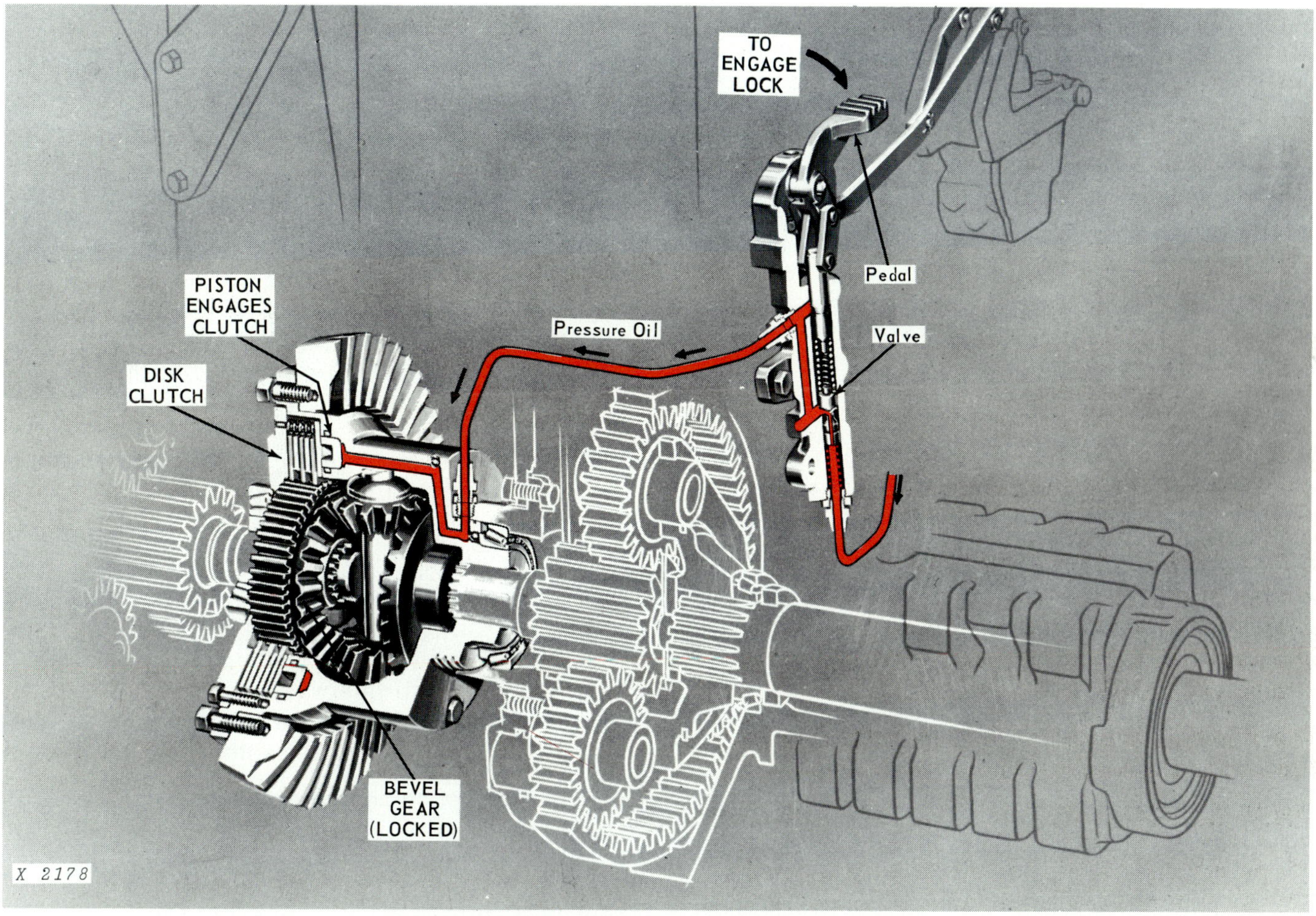

Fig. 4—Hydraulic Differential Lock On Farm Tractor

When the drive wheels again have equal traction, the lock disengages automatically. This happens because the collar is spring-loaded and releases once the torsional forces are gone.

This lock can be engaged "on the go." However, it should always be disengaged before turning the machine.

HYDRAULIC LOCK

The *hydraulic lock* (Fig. 4) uses pressure oil to lock out the differential.

When the operator depresses the pedal, the valve allows pressure oil to flow to the differential. The oil presses the piston against the disk clutch, compressing it against the housing. Since the disks are splined to the bevel gear and the plates are tanged to the housing, this locks up the bevel pinions and prevents differential action.

This lock can also be engaged "on the go".

When the operator releases the pedal, the valve cuts off oil pressure to the piston, allowing springs to return it, freeing the disk clutch. The differential then operates normally.

The pedal on this unit is linked to the tractor's brake pedals. This is a safety release and unlocks the differential whenever the brake pedals are used, preventing the danger of turning the tractor sharply with the differential lock still engaged.

 Litho in U.S.A.

AUTOMATIC (NO-SPIN) LOCK

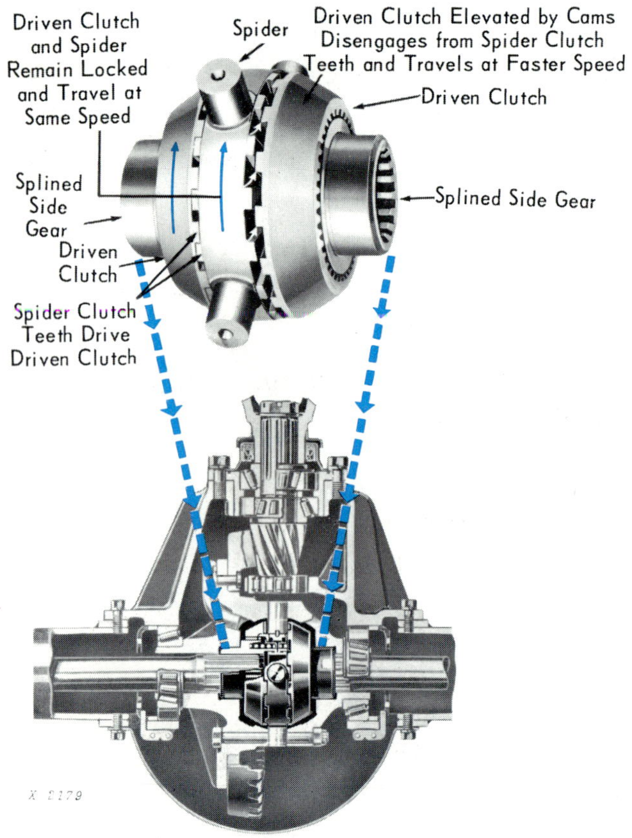

Fig. 5—Automatic (No-Spin) Differential Lock During A Left Turn

The *automatic (no-spin) lock* (Fig. 5) is different in that it is normally *engaged* and still allows for relative motion between the wheels on a corner.

A speed-sensitive differential control, when engaged, prevents one wheel from spinning when it loses traction and thereby depriving the other wheel of full driving power.

The no-spin unit is located inside the differential housing and replaces the bevel gears, bevel pinions, and their shafts.

Fig. 6—Driving Straight Ahead

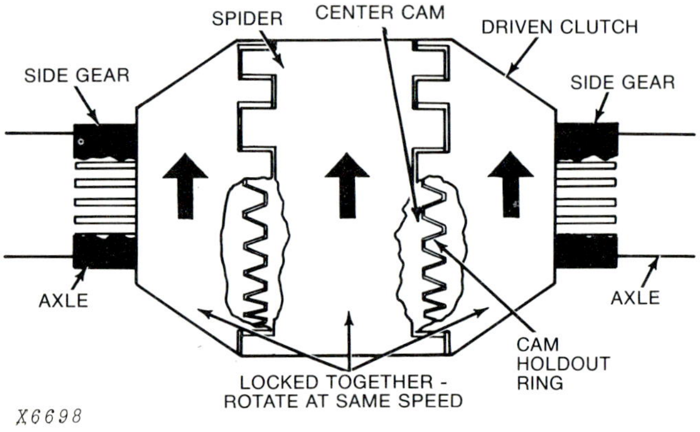

In Fig. 5, both axles are splined to the *side gears* of the no-spin unit—one on each side. The side gears in turn are splined to the *driven clutches.*

When driving straight ahead, the clutch teeth of the *spider*, and the driven clutches are fully engaged, and the spider drives the driven clutches. The whole system then rotates as a solid unit. The parts are held engaged by springs (Fig. 6).

When the machine is turned to the left the right driven clutch rotates faster to make the turn, through the overrunning action of the right wheel. The teeth on the right holdout ring slide up on the teeth of the center cam, forcing the teeth of the right driven clutch to disengage from the spider teeth (Fig. 7). Meanwhile, the left driven clutch is held engaged to the center cam by resistance of the slower left wheel.

As the turn is completed and the overrunning motion stops, the speed of the spider and the right driven clutch become the same again. There is a slight reversal of torque which causes the holdout ring to move back into phase with the teeth on the driven clutch, engaging all teeth again.

The whole mechanism then returns to the fully locked position until another turn.

The no-spin unit does not allow the inside wheel on a turn to slow down—only the outside wheel to speed up. For this reason, it cannot be used on machines with a great difference in wheel speeds on a turn. In a tight turn, these machines would not decelerate to a safe speed.

ADJUSTMENT OF DIFFERENTIALS

The ring gear and pinion in the differential are normally adjusted for bearing **preload** and gear **backlash.**

Chapter 1 has a story on "Adjusting The Gear Train." It explains the effects of wrong adjustment and gives an example of ring gear and pinion adjustment for a differential unit.

Fig. 7—Machine Turning Left

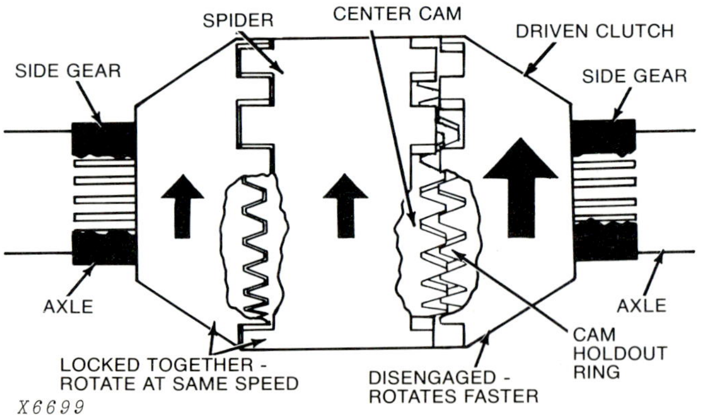

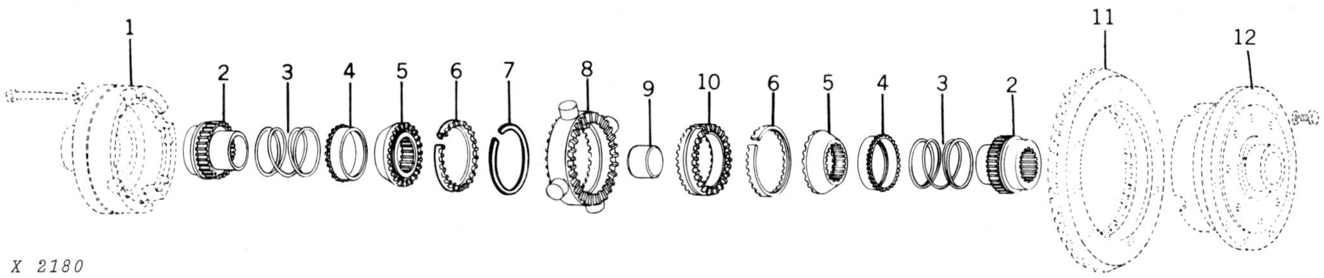

1 — Differential Housing Cover
2 — Side Gear (2 Used)
3 — Spring (2 Used)
4 — Spring Retainer (2 Used)
5 — Driven Clutch (2 Used)
6 — Cam Holdout Ring (2 Used)
7 — Snap Ring
8 — Spider
9 — Stop
10 — Center Cam
11 — Ring Gear
12 — Differential Housing

Fig. 8—Automatic (No-Spin) Differential Lock In Exploded View

TROUBLE SHOOTING

Here is a guide to the common failures of differentials and the possible causes.

1. NOISE IN DIFFERENTIAL AT ALL TIMES

a. Incorrect adjustment of ring gear and pinion.

b. Ring gear or pinion damaged.

c. Damaged bearings on pinion shaft.

d. Damaged bearings in differential housing.

2. DIFFERENTIAL NOT WORKING FREELY ON TURNS

a. Damaged or galled bearing surfaces between bevel gears and differential housing.

b. Damaged or galled bearing surfaces between bevel pinions and differential housing.

c. Damage or galling between bevel pinions and their shafts.

3. MECHANICAL DIFFERENTIAL LOCK DOESN'T HOLD

a. Linkage between engaging lever and shifting collar broken or improperly adjusted.

b. Shifting collar broken or missing.

4. HYDRAULIC DIFFERENTIAL LOCK DOESN'T HOLD

a. Too-low oil pressure.

b. Blown oil seals or gasket.

c. Failed control valve.

TEST YOURSELF

QUESTIONS

1. What *two* jobs does a differential do?

2. What are differential locks for?

3. Name the three types of differential locks.

4. Why is the hydraulic lock sometimes linked to the brake pedals?

5. Which type of differential lock is normally *engaged*?

6. True or false? "The no-spin differential allows one wheel to turn slower and the other faster than ring gear speed."

(Answers on page 10 in back of text.)

 Litho in U.S.A.

FINAL DRIVES / CHAPTER 8

Fig. 1—Final Drives In Modern Tractor

INTRODUCTION

The final drive is the last phase of the power train. It gives the final reduction in speed and increase in torque to the drive wheels.

Used on most large machines, the final drives are mounted near the drive wheels (Fig. 1).

By reducing speeds, the final drives lower the stress and simplify the transmission, since extra gears and shafts can be eliminated.

Most final drives must support the weight of the machine as well as withstanding torque and shock loads.

TYPES OF FINAL DRIVES

There are four major types of final drive systems:

- **Straight Axle**
- **Pinion**
- **Planetary**
- **Chain**

Each of these four drives have their advantages. All but the first give a speed reduction.

Let's look at each one in detail.

 Litho in U.S.A.

STRAIGHT AXLE DRIVES

The straight axle drive is the least complex of all final drives. It is commonly used in automobiles and light-duty trucks where no extra gear reduction for heavy loads and pulling power is required.

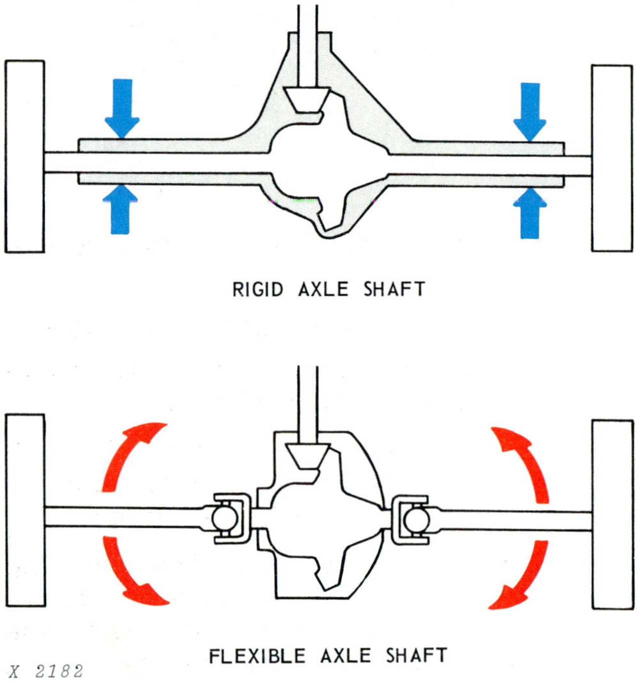

Fig. 2—Straight Axle Drives — Two Types

The drive wheels in a straight axle receive their power directly from the differential. For each revolution of the differential, the axle shaft and wheels make a revolution, giving a direct drive.

The drive axle in any final drive system is known as a "live" axle since it transmits the engine torque as well as supporting part of the machine weight. A "dead" axle on the other hand is one which only supports the machine, such as the front wheel axles of an automobile.

There are two types of straight axle construction:

- **Rigid Axle Shaft**
- **Flexible Axle Shaft**

The RIGID AXLE SHAFT (Fig. 2) is connected to the differential output by a splined coupling and is usually enclosed in an axle housing. The axle housing forms an intergral part of the differential housing, making one long housing from drive wheel to drive wheel.

The FLEXIBLE AXLE SHAFT (Fig. 2) is used when the drive wheels are independently suspended. The axles are connected to the differential by universal joints. This type of axle is sometimes referred to as a "swing axle". The drive wheel in this case is free to move vertically without affecting the position of the differential and transmission. Flexible axles are commonly found in compact automobiles and especially those with engines located in the rear.

FLOATING AXLE SHAFTS IN RIGID DRIVES

Two kinds of floating axles are being used today:

1) *Full-Floating Axles*
2) *Semifloating Axles*

They differ in whether or not they have to carry the weight of the vehicle in addition to turning the wheels.

Full-Floating Axle

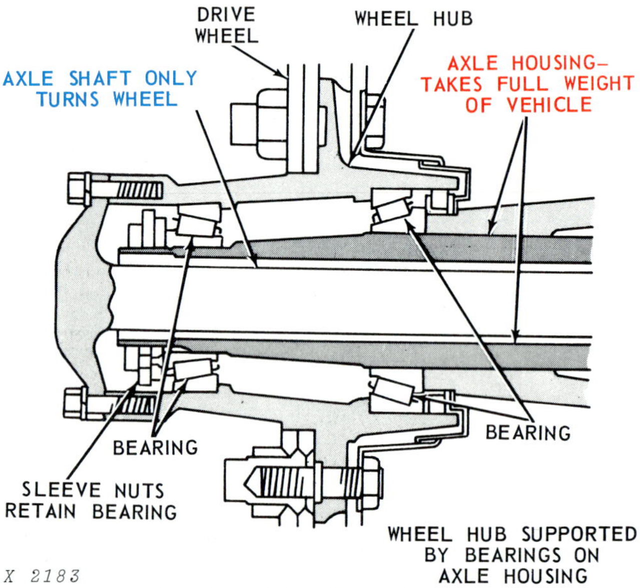

Fig. 3—Full-Floating Axle Shaft

The full-floating axle (Fig. 3) is used in many heavy-duty trucks. Each drive wheel is carried on the outer end of the axle housing on a pair of tapered roller or ball bearings. The bearings are positioned outside the axle housing. In this way the axle housing takes the full weight of the vehicle and absorbs all stresses or end thrust caused by turning, skidding, etc. The axle shaft then only transmits the torque from the engine and so "floats" in the axle housing. The shaft is connected to the drive wheel through a bolted flange and the axle shaft can be removed or serviced without removing the drive wheel.

 Litho in U.S.A.

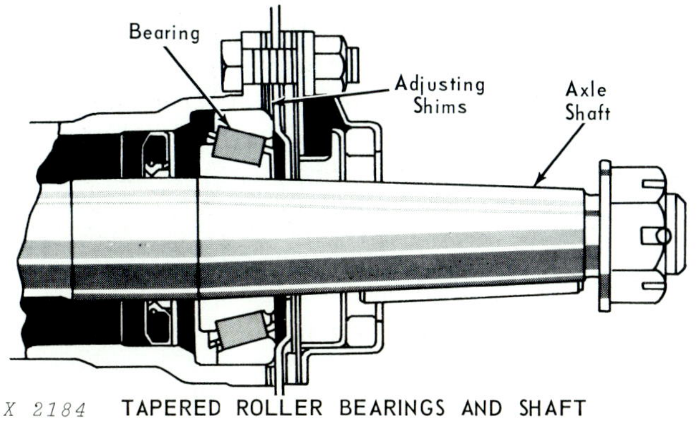

X 2184 TAPERED ROLLER BEARINGS AND SHAFT

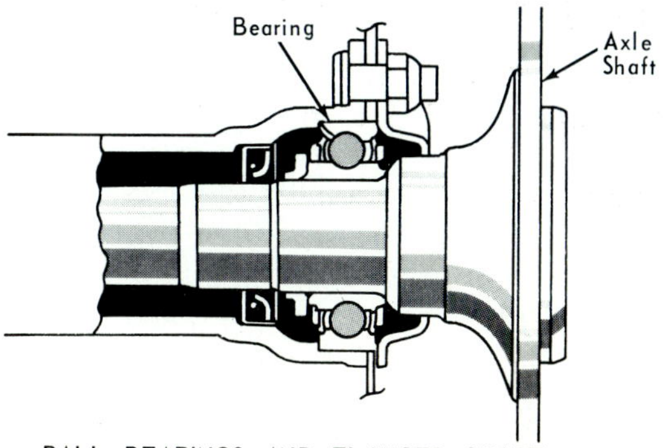

X 2185 BALL BEARINGS AND FLANGED SHAFT

Fig. 4—Semifloating Axle Shafts — Two Types

Semifloating Axles

Semifloating axle shafts (Fig. 4) float in the differential in the same way as full-floating axles. The main difference is at the outer ends of the axle housing. Here a single bearing assembly is positioned between the axle shaft and axle housing (Fig. 4). Therefore, the shaft supports the weight of the vehicle as well as transmitting the engine torque. The axle shafts must now take all stresses caused by turning, skidding, etc.

Two types of wheel bearings are used in semifloating axles—either tapered roller bearings or ball bearings (Fig. 4).

When tapered roller bearings are used, an adjustment for axle shaft endplay is provided. This is done with shims or an adjusting nut at one end of the axle housing. In order to transmit shaft end thrust equally to the bearings at both ends of the axle housing, a thrust bloc, or spacer is located between the inner ends of the two axle shafts within the differential carrier.

Semifloating axles are commonly used in automobiles and light trucks. The outer ends of the shaft may be flanged or tapered as shown in Fig. 4. When the shaft is tapered, the wheel hub is keyed to the shaft and locked in place with a nut.

PINION DRIVES

A spur gear and pinion are used for the final gear reduction on many tractors and self-propelled machines. This enables the differential to transmit power at higher speeds and lower torque.

There are two types of pinion and gear reductions:

- **Pinion drive located within differential case.**
- **Pinion drive located on outer ends of final drives.**

Each has a purpose and some advantages, as we'll see now.

PINION DRIVE WITHIN DIFFERENTIAL CASE

The advantages of having the gear reduction directly driven from the differential are:

1) All gears are enclosed within the differential-transmission case. This is a more compact unit and only one lubrication system is needed.

2) The final drive is in a straight line, permitting a straight axle shaft. On farm tractors this is a big advantage, since a straight axle provides a large range of wheel tread adjustment for operating in various row-crop spacings.

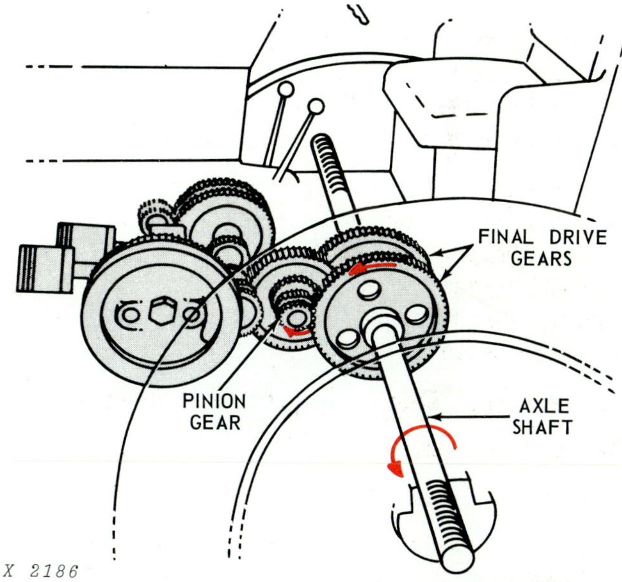

X 2186

Fig. 5—Pinion Gear Reduction Within Differential Case

Power is transmitted to the drive wheels through pinion gears which are connected to the differential output. The pinions mesh with the larger final drive gears which in turn rotate the axle shaft (Fig. 5).

The axle shaft is carried on two tapered bearings, while the final drive gear is splined to the inner end of the shaft. The bearings are positioned internally at both ends of the shaft between the axle housing and shaft. Bearing adjustment is usually made with an adjusting nut. However, some final drives may be equipped with a shim adjustment.

The axle supports the weight of the tractor and absorbs end thrust, as well as transmitting the engine torque.

PINION DRIVE ON OUTER ENDS OF FINAL DRIVE

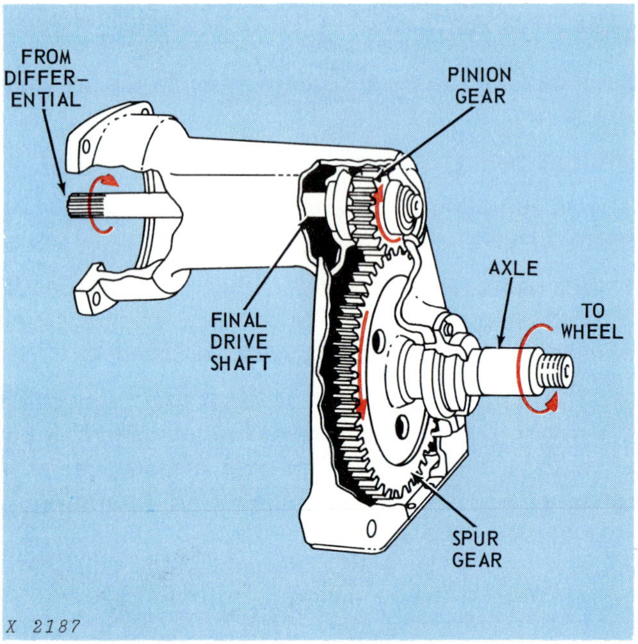

Fig. 6—Pinion Gear Reduction On Outer Ends Of Final Drives

Placing the pinion and gear reduction drive at the outer end of the final drive gives more clearance under the axle. The gears are contained in a separate case or housing (Fig. 6) and have their own oil reservoir. Power flow is shown by the red arrows.

This system is popular for good crop clearance in farm tractors and other machines.

On most farm tractors, the drive shaft is enclosed in a housing which is attached to the differential case. In turn the gear reduction housing is attached to the end of the drive shaft housing, forming the final drive assembly. This final drive construction is sometimes called a "drop axle housing". In this case, the entire final drive assembly carries the weight of the tractor.

PLANETARY DRIVES

The planetary gear reduction system is smaller and more compact than the pinion type. Planetary systems are also more durable under heavy loads since torque loads are spread more evenly over several gears.

The use of planetary gear systems is becoming more common on farm tractors, industrial machines, and trucks as engine horsepower and pulling loads are increased.

The planetary gear set may be located next to the differential (Fig. 7) or at the outer ends of the final drive (Fig. 8). Both types—inboard and outboard—operate the same, as described below.

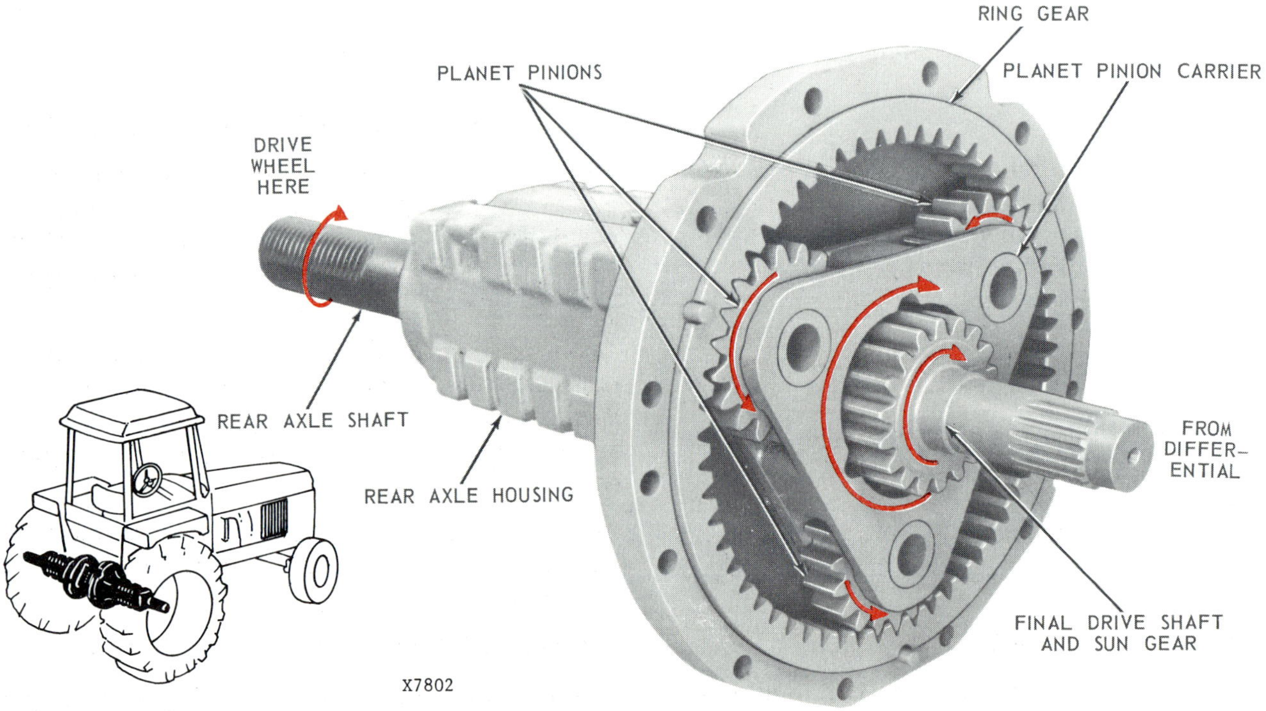

Fig. 7—Final Drive With Planetary Gears Next To Differential (Inboard Type)

 Litho in U.S.A.

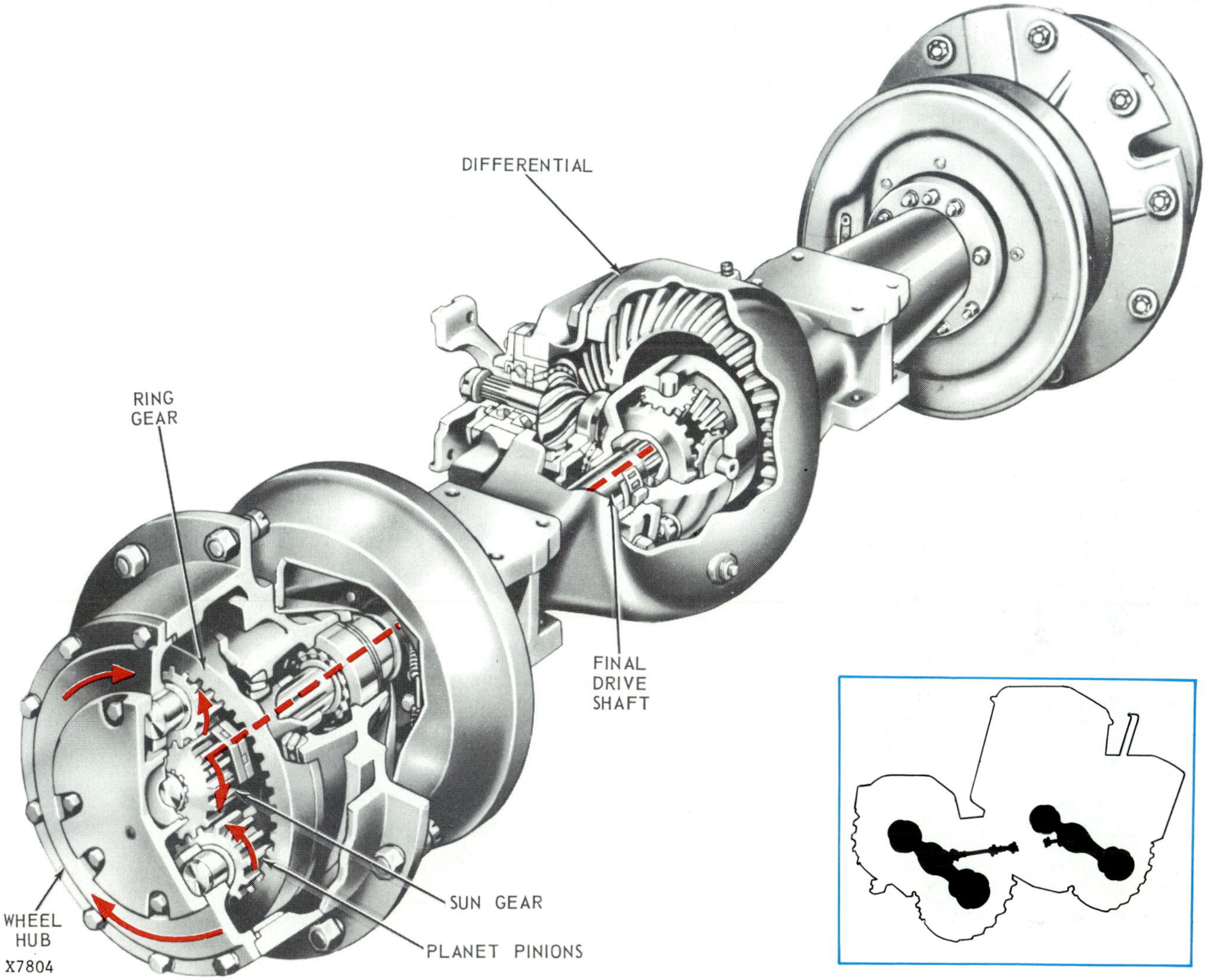

Fig. 8—Final Drive With Planetary Gears At Outer Ends (Outboard Type)

OPERATION

The planetary gear reduction system receives engine power from the differential through the final drive shaft and sun gear. The sun gear is an integral part of the final drive shaft and so turns with the shaft. The sun gear meshes with the planet pinions which are mounted in the planet pinion carrier.

As the sun gear turns, it forces the planet pinions to "walk around" the inside of the ring gear as shown by the arrows. The planet pinion carrier is forced to rotate in the same direction as the sun gear, thus delivering engine power to the drive wheels *but at a reduced speed and an increased torque.*

CONSTRUCTION OF PARTS

Let's look at the construction of the planetary systems shown in Figs. 7 and 8.

Planetary Mounted Next To Differential (Fig. 7)

With this planetary, a straight, in-line axle is achieved. This permits a large range of wheel tread adjustments on farm tractors for various widths of crop rows. In addition, all gears are located in one compact housing.

The axle shaft is carried on two tapered bearings with the inner end of the shaft splined to the planet carrier (Fig. 7). The bearings are positioned internally at both ends between the axle housing and shaft.

With this type of axle, the axle shaft supports the weight of the tractor and absorbs end thrust, as well as transmitting engine torque.

Planetary At Outer Ends Of Final Drives (Fig. 8)

When the planetary is located at the outer ends of the final drive, the final drive system is enclosed in one rigid housing (Fig. 8). That is, the differential housing, both axle housings, and the planetary gear housings (shown) are connected to form a one-piece construction.

The axle shaft in this system is the *full-floating* type which was described earlier. The only difference is that the planetary carrier drives the wheel hub instead of a flanged end of the axle shaft.

CHAIN FINAL DRIVES

The chain final drive system (Fig. 9) provides a greater clearance under the axle than any other system. An example of this is in high-clearance self-propelled sprayers which must operate in high, standing crops.

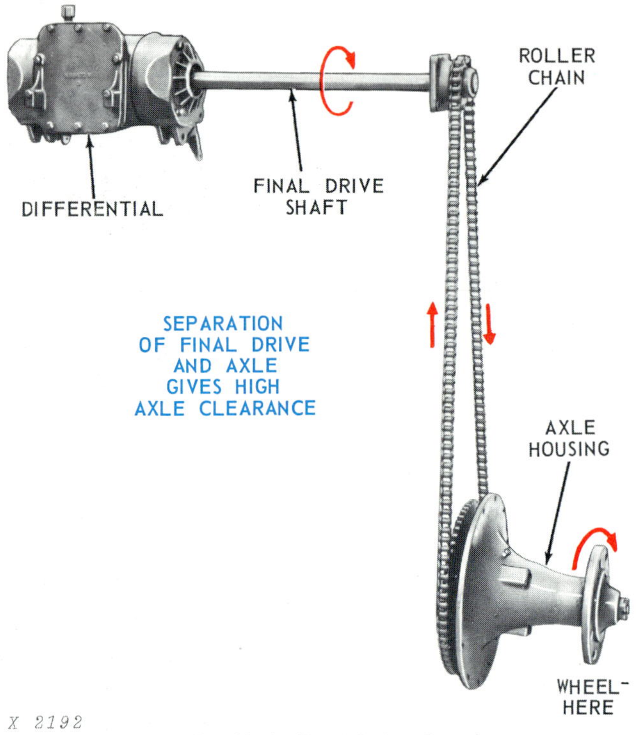

Fig. 9—Chain Final Drive System

The disadvantage of a chain drive is its tendency to loosen due to wear and stretching. Wear can be held to a minimum by having the chain operate in an oil bath. In most cases, provisions are made to take up the chain slack.

In Fig. 9, an eccentric bearing is provided in the axle housing so that the chain slack can be taken up by rotating the bearing housing. In some tractors, the entire axle housing can be moved for chain adjustment.

Another method of adjusting the chain is the use of a tightener sprocket on the slack side of the chain. However, when the machine is reversed, extra force is exerted to the slack side of the chain and through the tightener sprocket.

For details on roller chains, see Chapter 10.

MAINTENANCE OF FINAL DRIVES

The reliability of any final drive depends upon good maintenance, operating at rated load, and proper repair once a failure has occurred.

Watch these key points when diagnosing final drive failures:

(1) Excessive drive shaft endplay

(2) Overheating

(3) Lack of lubrication

EXCESSIVE DRIVE SHAFT ENDPLAY

Excessive drive shaft endplay is normally caused by loose shaft bearings. However, it can be created by:

(a) Foreign material in the lubricant which will wear bearings rapidly;

(b) Overloading the machine either by weight or engine torque;

(c) Poorly adjusted bearings at the time of assembly.

A continuous noise or knock is a good sign of loose or damaged bearings. On machines with semifloating axles, such as automobiles, the knocking noise can be heard in the differential case, since the ends of the axle shaft are rapping the spacer block. Readjustment of the bearings that are worn or damaged will not provide a satisfactory repair. Instead, replace the bearings.

Compare the roller ends of a new tapered bearing with a worn bearing. Worn bearing rollers have no shoulder compared to a new bearing—replace them.

OVERHEATING

Many final drives are damaged simply by overheating. This is caused by not maintaining the lubricant at the proper level or by using the wrong type of lubricant.

Galling, pitting, or scoring on the surface of a part indicates the lack of lubricating film and that overheating has occurred.

 Litho in U.S.A.

Overloading and abuse of the machine will also cause overheating. Excessive loads cause deflection in the final drive assembly and concentrate stresses and friction in one area.

LACK OF LUBRICATION

Loss of lubricant through worn and broken oil seals and gaskets may cause severe damage to the final drive. While some bearings are automatically lubricated by oil creeping along the drive shaft from the differential, others are sealed off and require separate lubrication.

When installing a new oil seal, make sure it is flexible or pliable in the area where the seal fits around the shaft. Soak the seal in oil before installing it.

ADJUSTING FINAL DRIVES

To prevent premature failures, adjust the final drives properly. This includes adjusting the axle bearings—either preloading or allowing a specified endplay.

ADJUSTING AXLE BEARINGS

Loose gears or bearings that are too tight or too loose will break down prematurely, regardless of their strength and design.

Bearings must be adjusted to the manufacturer's specifications. Some designs require the bearings to be preloaded, while others require a slight amount of shaft endplay.

Preloading Bearings

When the manufacturer's instructions call for preloading the bearings, bearings must be placed under slight tension. In some cases, this is done by SLIGHT overtightening of the bearing adjusting nut, while in other designs it is done with shims. Preload is normally measured in inch-pounds of rolling or rotating torque (NOT starting torque).

Some torque specifications will include frictional drag such as oil seals, gears, etc. in the final drive. If it is not included, measure the rolling torque while the axle shaft still has a slight amount of endplay, and *add* this torque reading to the specified torque. Then make the necessary bearing preload adjustment. Also, when new bearings are installed, a greater amount of torque will be required. Always make the preload adjustment in small increments until the proper preload is obtained.

There are several methods of measuring rolling torque.

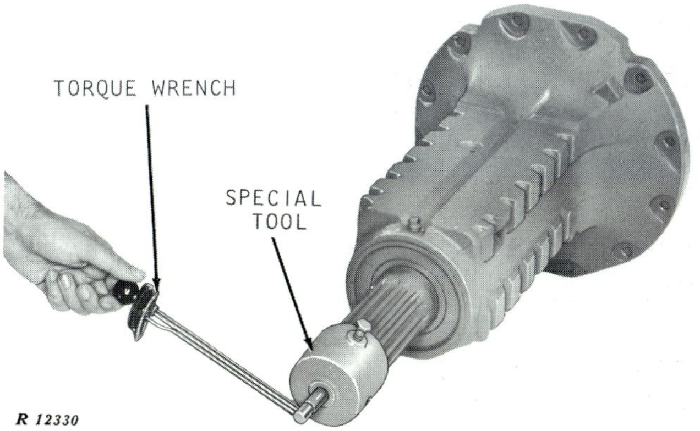

Fig. 10—Checking Bearing Preload With A Torque Wrench

One method is to use a special tool or adapter with a torque wrench (Fig. 10). This gives a direct reading on the wrench.

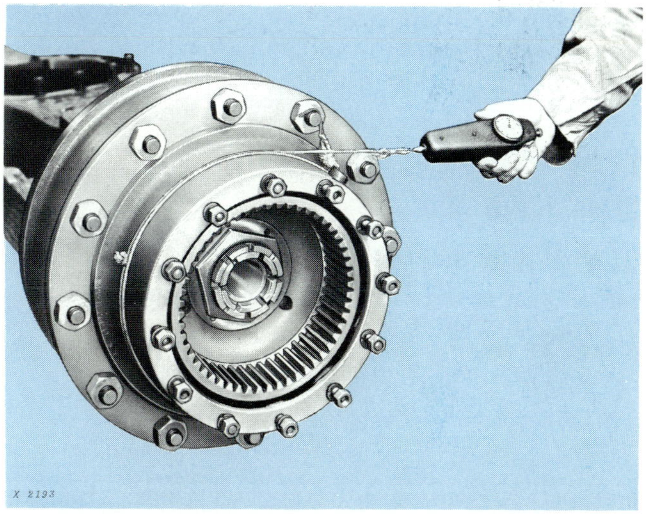

Fig. 11—Checking Bearing Preload With Pull Scale

Another method of checking the preload torque is by means of a cord and pull scale (Fig. 11). The preload is figured by multiplying the radius (distance from the center of the drive shaft to a point on the circumference from which the cord is pulled) by the reading on the pound scale.

For example: Assume the radius is 4 inches (100 mm) and the reading on the scale is 7 pounds (31 N). Multiply 4 inches (100) by 7 pounds (31) and you will have 28 inch-pounds (3.1 N•m) of rolling torque (100x31 = 3100 N•mm = 3.1 N•m).

Adjusting For Endplay

Some types of drive axles call for the axle shaft to have a slight amount of endplay. This means that the shaft when properly adjusted must be free to move endways within the limits specified. The end movement in most cases is between 0.001 and 0.010 inch (0.025 and 0.25 mm).

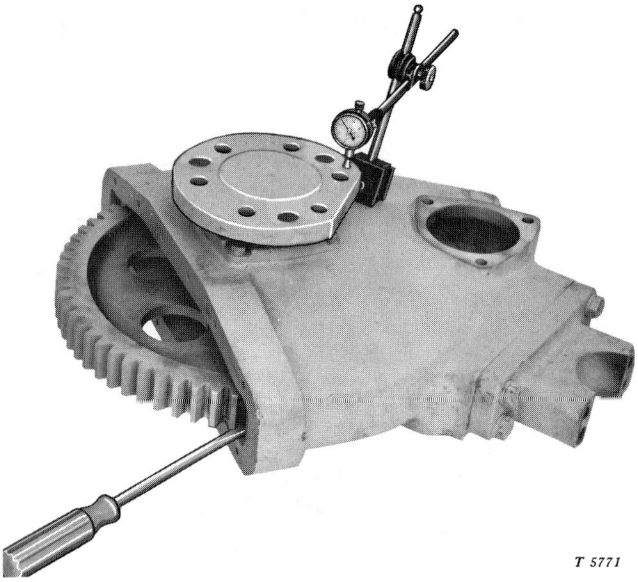

Fig. 12—Checking Final Drive Shaft Endplay With A Dial Indicator

To obtain the proper endplay, shims or an adjusting nut are provided. Measure the endplay with a dial indicator (Fig. 12).

TEST YOURSELF

QUESTIONS

1. What are the four major types of final drive systems?

TEST YOURSELF

QUESTIONS

1. What are the four major types of final drive systems?

2. Which one does *not* give a speed reduction?

3. What is the difference between a "live" and "dead" axle?

4. True or false? "In a semifloating axle, the weight of the machine is not carried on the axle shaft."

5. Name one advantage of the planetary final drive over the other three types.

6. What is the main advantage of the chain final drive? The main disadvantage?

(Answers on page 10 in back of text.)

POWER TAKE-OFFS / CHAPTER 9

X2118

Fig. 1—Power Take-Off (PTO) For Modern Tractor

The power take-off (PTO) is an attachment in the power train of a machine. It drives auxiliary equipment.

TYPES OF PTO'S

There are three basic types of PTO's:

- **Transmission-Driven**
- **Continuous-Running**
- **Independent**

The three types are compared in Fig. 2.

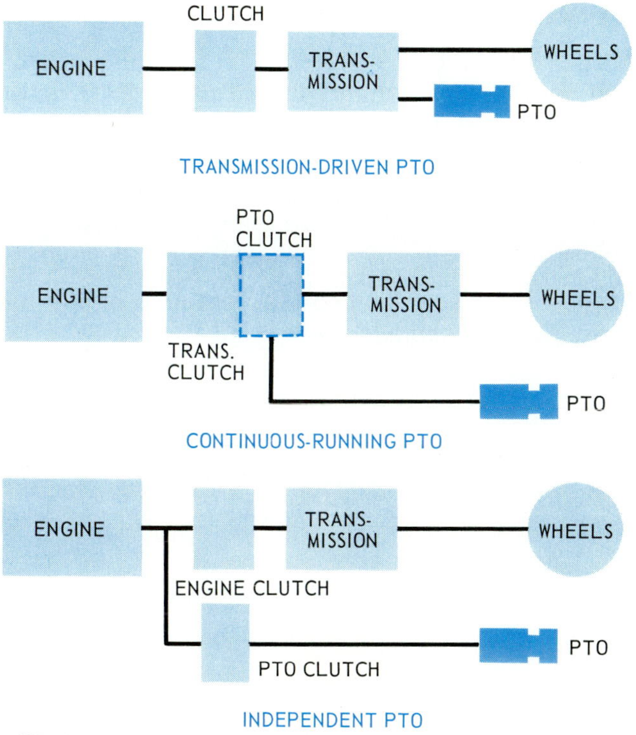

X3314

Fig. 2—Three Types Of Power Take-Offs (PTO'S)

TRANSMISSION-DRIVEN PTO's operate only when the engine clutch is engaged. They stop whenever the clutch is disengaged (stopping machine travel).

CONTINUOUS-RUNNING PTO units have two clutches—one for the transmission and one for the PTO. Both clutches are operated by one control. The first half of the control travel operates the transmission clutch, while the second half operates the PTO clutch. Therefore, the PTO can be operated with the transmission disengaged (machine travel stopped). However, the transmission cannot be engaged again while the PTO is operating.

Litho in U.S.A.

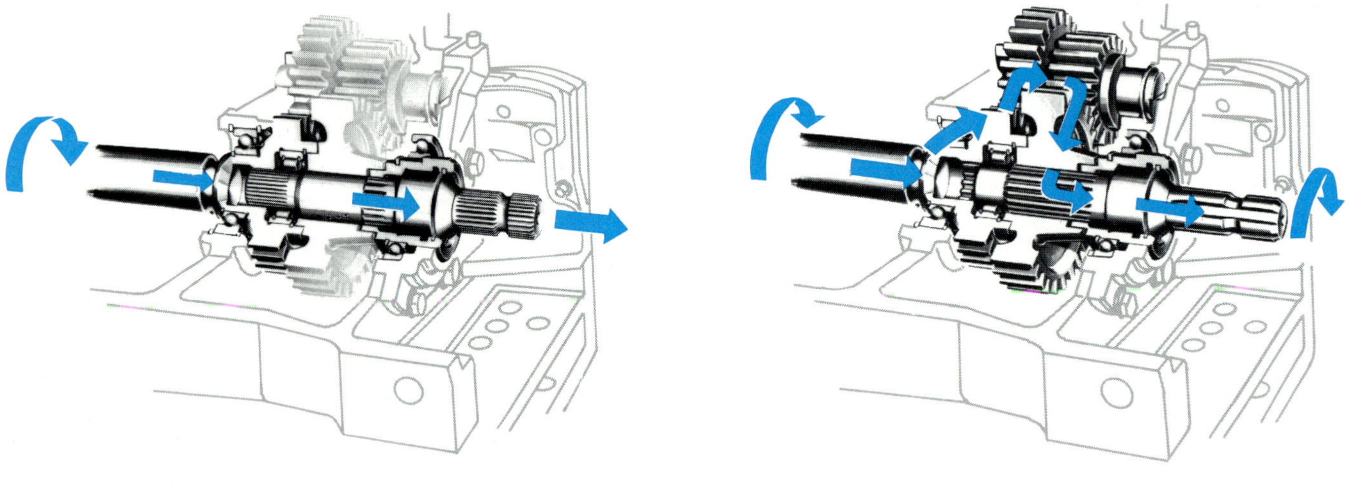

X2120 PTO OPERATING AT 1000 RPM PTO OPERATING AT 540 RPM

Fig. 3—Operation Of Power Take-Off (Dual Speed Type)

INDEPENDENT PTO's have their own clutch controls completely separate from the engine clutch and transmission. This means that the PTO can operate while the machine travel is stopped and also that the PTO can be disengaged and engaged while the machine is in motion (not possible with continuous-running PTO's).

OPERATION OF PTO

Most PTO's are gear-driven from the transmission and send power through a shaft to the PTO outlet where the driven equipment is coupled.

Let's look at the operation of a typical PTO.

One PTO (Fig. 3) offers two speeds—540 rpm and 1000 rpm.

To get these speeds, two different stub shafts are used as we'll explain now.

OPERATION AT 1000 RPM

Operation at 1000 rpm is shown at left in Fig. 3.

The 1000 rpm stub shaft is installed in the PTO outlet so the splines engage with the splines on the inside of the PTO shaft. Engine power is transmitted directly through to the PTO stub shaft.

The engine is set at a certain speed (such as 2100 rpm) to get the speed of 1000 rpm at the PTO.

OPERATION AT 540 RPM

Operation at 540 rpm is shown at right in Fig. 3.

A different stub shaft is installed in the PTO outlet for 540 rpm operation. This shaft engages a new set of splines inside the PTO drive gears, so power is transmitted through the upper idler gears as shown by arrows. The speed is reduced to 540 rpm by the gear ratios.

PTO DRIVES

Once the power leaves the PTO outlet, it must be transmitted to the driven equipment. This is the job of the PTO **drive** (Fig. 4).

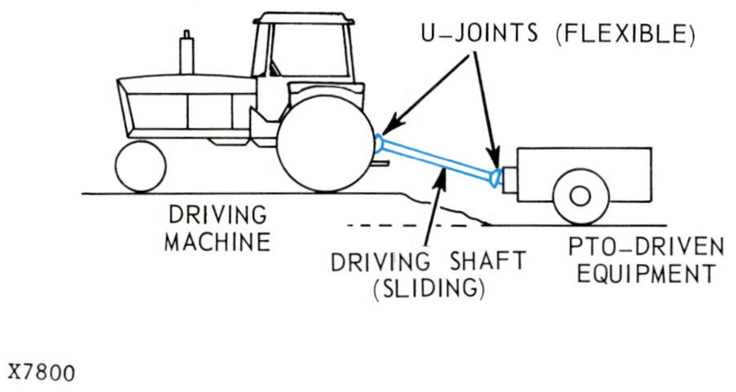

Fig. 4—PTO Drive

The PTO drive must be flexible for most equipment, especially if it is pulled by the driving machine. For these PTO drives, two major parts are needed:

1) Universal Joints

2) Drive Shaft

Let's look at each part in detail.

UNIVERSAL JOINTS

A universal joint is necessary to join two rotating shafts in different planes (Fig. 4).

The most common type is the Cardon or **Hooke joint** (Fig. 5).

This joint consists of two U-shaped yokes fastened to the ends of the two shafts to be connected. Inside these yokes is a cross-shaped journal which holds the yokes together and allows each yoke to bend, or pivot, with respect to the other. This allows one shaft to drive the other at angles up to 30 degrees out of alignment.

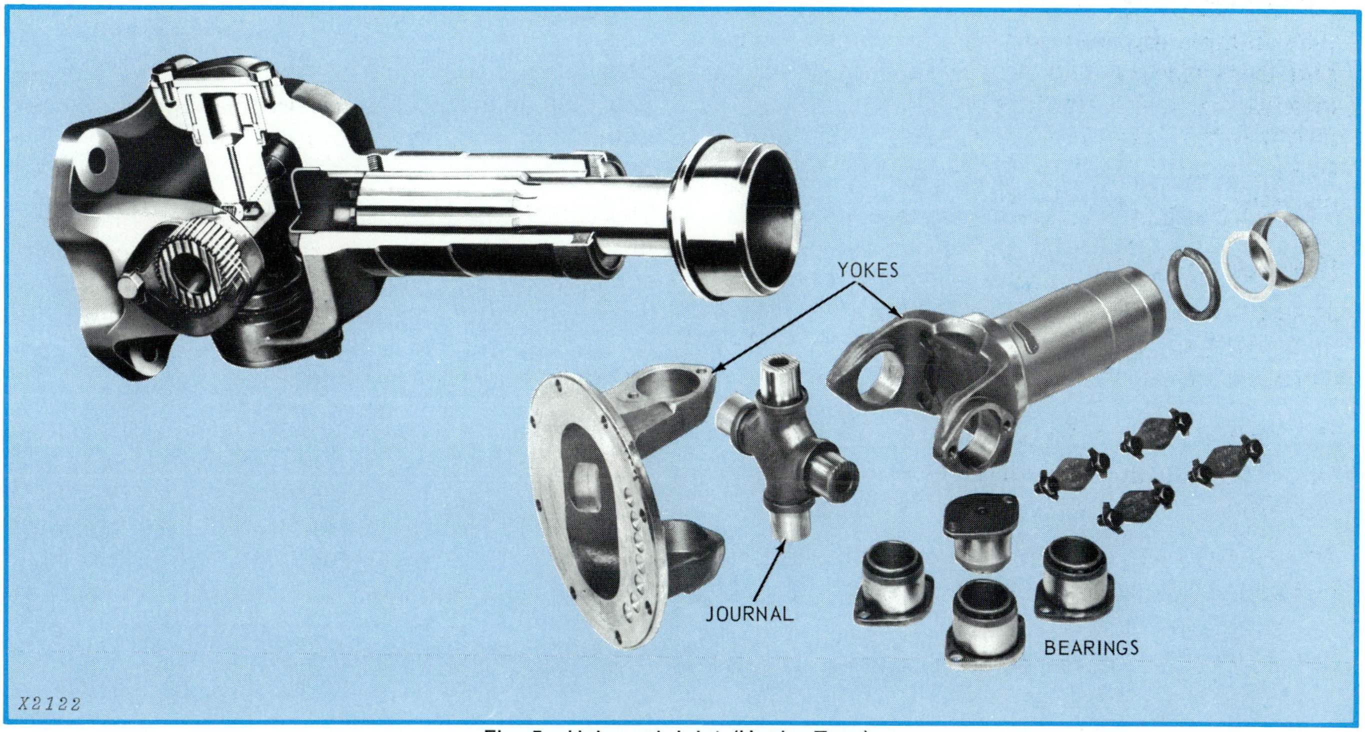

Fig. 5—Universal Joint (Hooke Type)

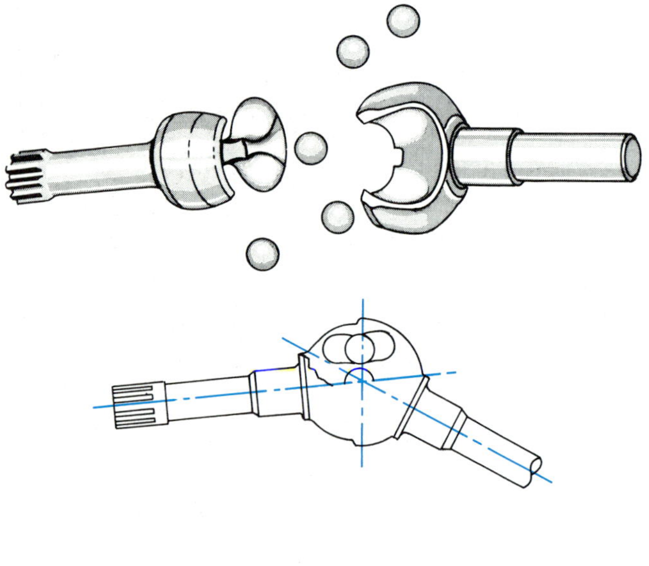

Fig. 6—Universal Joint (Bendix-Weiss Type)

Another type of universal joint is used to transmit a smoother torque when less power is needed. This type of universal joint is the **Bendix-Weiss** (Fig. 6). This type of joint has four large balls to transmit the rotary force, with a small ball as a spacer.

The Hooke universal joint gives a whipping motion as the shaft rotates. The amount of whipping depends on how much the shafts are out of alignment. The Bendix-Weiss joint transmits rotary motion at a constant rate, but this joint is more expensive and has less strength than the Hooke type.

Most farm and industrial machines use the Hooke type of U-joint.

DRIVE SHAFTS

There are two basic types of drive shafts:

- *Solid shaft*
- *Telescoping shaft*

The difference is that one shaft has a flexible length while the other does not.

Solid Shafts

The solid shaft has a fixed length and a universal joint at each end. The distance between the two joints is always fixed. This type of drive is used when power is to be transmitted from one fixed plane to another.

Telescoping Shaft

The telescoping shaft has a sliding element and can have either two or three universal joints.

The telescoping shaft with *two* universal joints has one at the machine end and another at the equipment end (Fig. 7). Telescoping occurs between these two points at the slip joint.

The telescoping shaft with *three* universal joints has one solid PTO shaft (on a mount), one telescoping shaft, and three universal joints. This gives the flexibility of the two-joint type, yet allows for the change in hitch angle when equipment is converted to transport position.

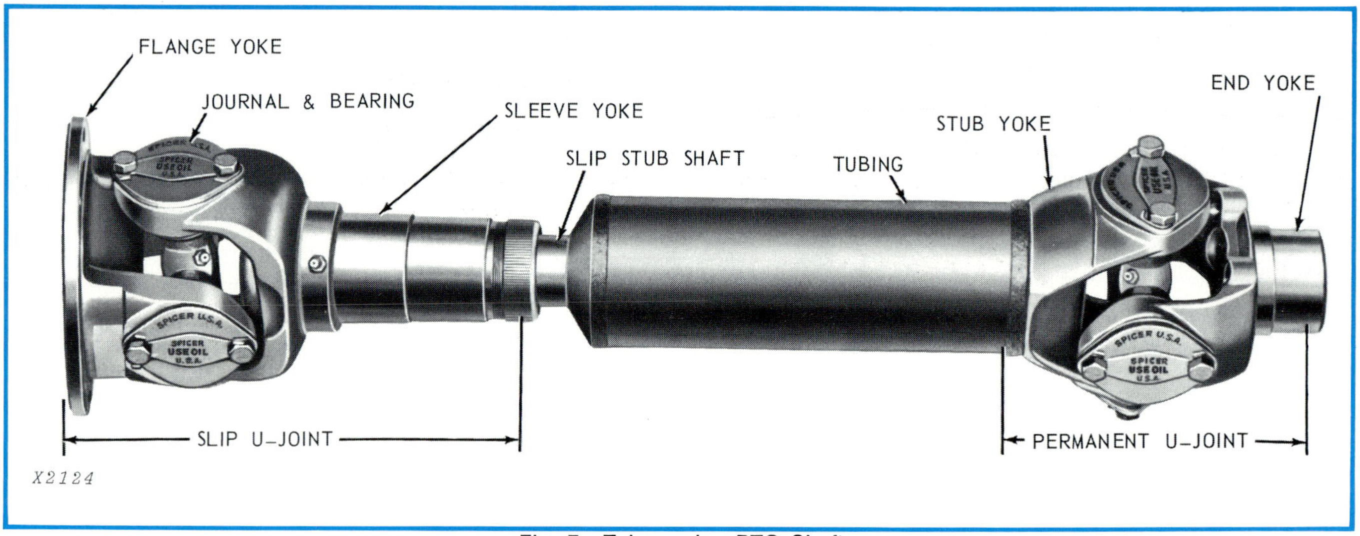

Fig. 7—Telescoping PTO Shaft

STANDARDS FOR PTO DRIVES

ASAE-SAE Tractor Standards

The American Society of Agricultural Engineers (ASAE) along with the Society of Automotive Engineers (SAE) have worked for years to establish standards for tractor and implement manufacturers. These standards help in changing equipment from one tractor to another.

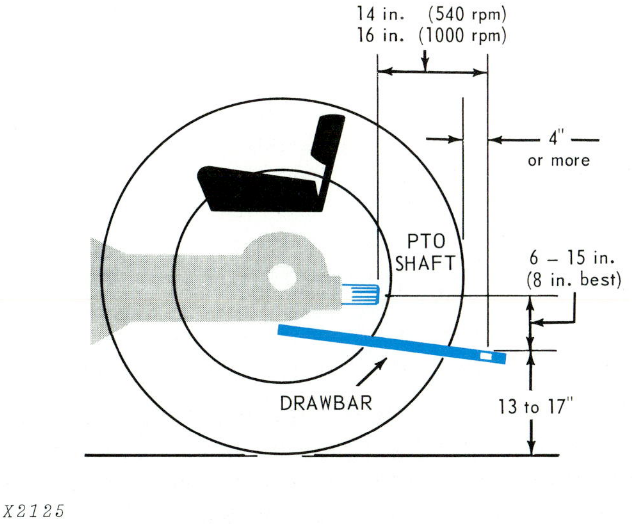

Fig. 8—Location Of PTO Shaft And Drawbar Hitch On Farm Tractor (ASAE-SAE Standards)

The recommended location of the drawbar hitch point in relation to the PTO shaft for equipment drawn behind a tractor is shown in Fig. 8.

In addition, the lateral location of the PTO shaft should be within 3 inches of the tractor centerline, and the hitch point should be directly under the extended centerline of the PTO shaft.

The PTO shafts of tractors have also been standardized. The shafts have either six splines and rotate at 540 rpm, or twenty-one splines and rotate at 1000 rpm (Fig. 9).

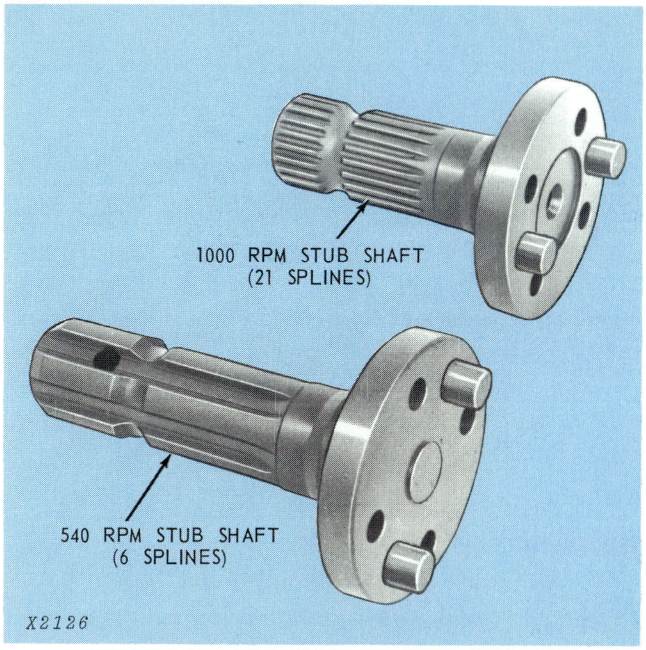

Fig. 9—PTO Shafts—Two Types

All PTO shafts rotate clockwise when viewed from the rear of the tractor.

PTO Shielding Standards

The tractor PTO shaft and the equipment drive shaft must be covered at all times.

The tractor PTO shaft shield is provided by the tractor manufacturer, while the equipment drive shaft shielding is provided by the equipment manufacturer.

The ASAE and SAE have standards by which manufacturers can build master shields for their PTO shafts.

The equipment drive line has two basic types of shields:

1) The tunnel shield

2) The spinner shield

The tunnel shield is merely an inverted U-shaped piece of steel which covers the moving parts.

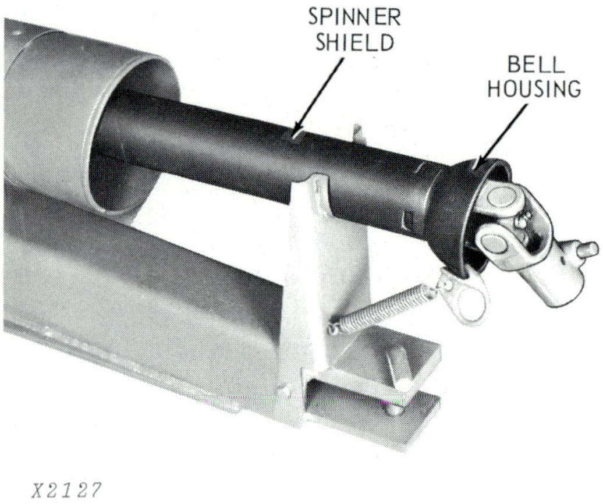

Fig. 10—Spinner Shield For PTO Drive Shaft

The spinner shield (Fig. 10) is a cylinder-shaped piece which surrounds the shaft with a bell housing on each end to cover the U-joints.

The spinner shield rotates with the shaft except when contact is made with a foreign object. When this happens, the spinner shield stops rotating and allows the shaft to rotate freely inside the shield.

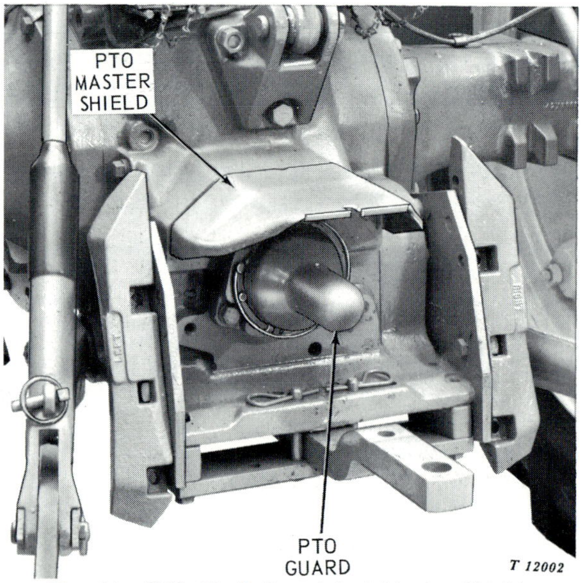

Fig. 11—PTO Shaft Guard And Master Shield

SERVICING THE PTO DRIVE

The service life you get from your PTO drive depends on the service you give it.

Lubricate the drive line in three areas:

- **The telescoping drive shafts**
- **The bearings**
- **The universal joints**

The lubrication intervals will vary with the field or job conditions. Consult the machine operator's manual.

SAFETY RULES FOR PTO'S

Safety is a prime factor in equipment design.

Here are safety rules for working on machines equipped with PTO's.

1. Keep PTO guards and shields in place even when the PTO is not operating.

2. Always disconnect the PTO when not in use.

3. Never engage the PTO while the machine engine is shut off.

4. Keep hands, feet, and clothing away from PTO parts.

5. Be sure your clothing is relatively tight and belted. Loose clothing can become caught in moving PTO parts.

6. Never operate PTO shafts at extreme angles.

7. Never ride, or permit others to ride, on the drawbar of the tractor.

8. Never hook 540 rpm equipment to a 1000 rpm PTO or vice-versa.

9. Be sure PTO spinner shields rotate freely at all times. Disengage all power and shut off tractor engine before checking spinner shields.

10. Always be sure PTO drive shaft is properly secured to the machine's PTO shaft.

11. Do not service, lubricate, or perform other work on a machine without first disengaging the machine PTO and shutting off the engine.

Fig. 12—Failures Of U-Joint Journals

TROUBLESHOOTING PTO'S

Problem	Possible Cause
1. Machine vibrates excessively while operating	PTO shaft is not properly aligned
2. Twisted PTO shaft	Overload on PTO shaft
	Improper hitch setting
3. PTO shafts not telescoping properly	Worn bearings
	Spinner shields or shaft rusted from poor lubrication
4. Brinnelling of U-joint journal (Fig. 12)	Excessive metal fatigue
	Poor heat treatment
5. Galling of U-joint journal cross ends (Fig. 12)	Drive shaft rpm too high
6. Journal cross end and cups chipping	U-joint capacity too low for required hours and angle
7. Abrasive corrosion on PTO shaft	Extreme low angle operation

TEST YOURSELF

QUESTIONS

1. Match the three items at left below with the correct definition at right.

a. Independent PTO	1. Operates only when main engine clutch is engaged, stops when disengaged.
b. Continuous-running PTO	2. Completely separate from main engine clutch.
c. Transmission-driven PTO	3. Can be operated even though main power train is disengaged.

2. How can you tell two PTO stub shafts—540 and 1000 rpm—apart?

3. True or false? "Never engage the PTO while the machine engine is shut off."

(Answers on page 10 in back of text.)

SPECIAL DRIVES / CHAPTER 10

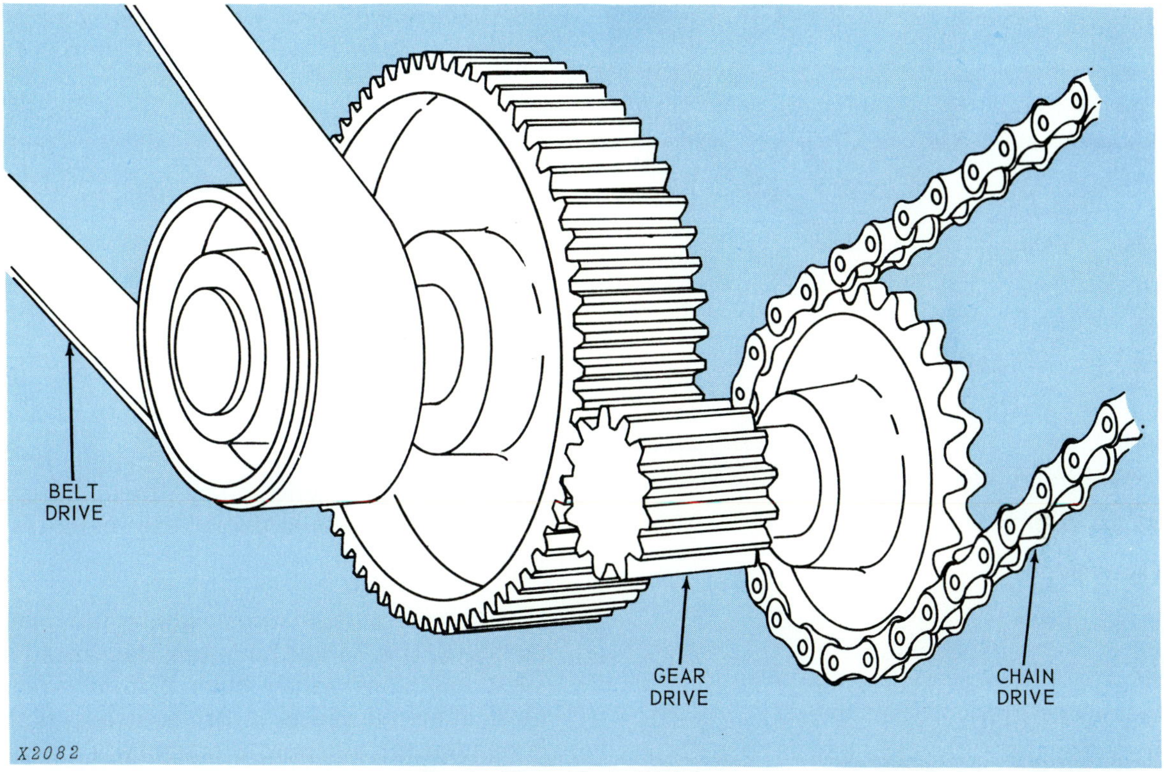

Fig. 1—Three Kinds Of Drives

In this chapter, we will cover these special drives not included in other chapters of this manual:

- **Belt Drives**
- **Chain Drives**
- **Gear Drives**

Refer to Fundamentals of Service — Belts and Chains for more information on these subjects.

Fig. 1 shows these three drives in operation.

We will also look at reciprocating drives and at safety devices like slip clutches, jump clutches, and shear pins.

BELT DRIVES

Belt drives are *frictional* drives. They transmit power by contact between the belt and the driving and driven pulley. Belts should have a firm but not a sticky grip on the pulleys. One pulley in any belt drive should be adjustable.

The ability of belts to transmit power depends on:

1. *Tension holding the belt to the pulleys.*
2. *Friction between belt and pulleys.*
3. *Arc of contact (or "wrap") between belt and pulleys.*
4. *Speed of belt (higher speeds cut tension and contact).*

Belts are normally used to transmit power between two parallel shafts but they can also be used in a variety of other ways (Fig. 3).

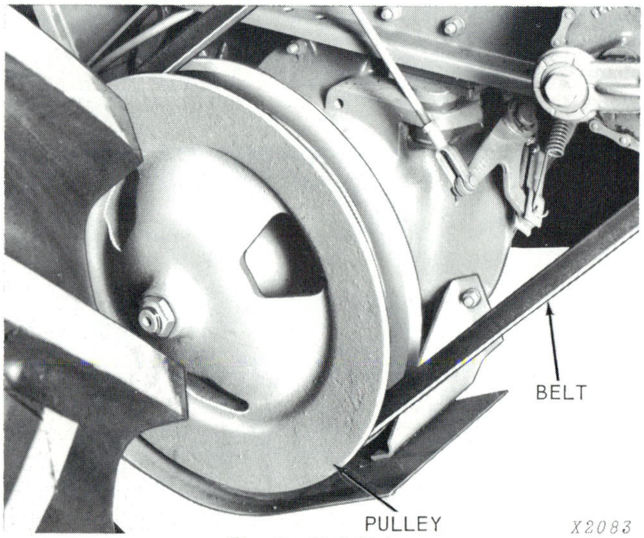

Fig. 2—Belt Drive

ADVANTAGES OF BELT DRIVES

1. *Easy to design*
2. *Inexpensive*
3. *Absorb shock*
4. *Easy to install*
5. *Easy to service*
6. *Permit simple power cutoff*

DISADVANTAGES OF BELT DRIVES

1. *Do not last as long as some drives*
2. *Will not carry as heavy a load*
3. *Cannot be used where exact timing is required (unless a special tooth-type belt is used)*

TYPES OF BELTS

Belts are of three basic types:

- **Flat Belts**
- **V-Belts**
- **Round Belts**

Belts may be constructed of leather, fabric, rubber, rayon, neoprene, hemp rope, thin flat steel, wire, and combinations of these materials.

Flat Belts

Flat belts are used most where the pulleys are far apart. Use of flat belts has decreased because most machines today have built-in drives. Their most common use is in drives where high power is needed for a separate machine. *Examples:* old-time threshers, sawmills.

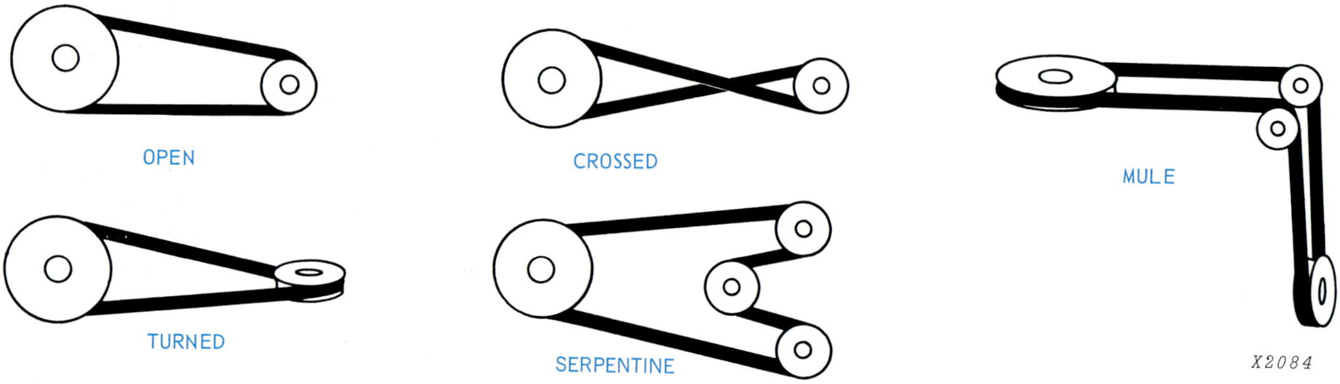

Fig. 3—Forms Of Belt Drives

 Litho in U.S.A.